W9-BZG-426
"Engrossing." —The Washington Post
"Beautifully written." —Time
"A gripping biography, written with intelligence, warmth, and panache." —Steven Pinker

"An illuminating glimpse into some of psychology's shamefulness, as well as one of its greatest successes." —*The Dallas Morning News*

"Engrossing . . . remarkably evenhanded." —*The Cleveland Plain Dealer*

"Enormously interesting . . . As a piece of science history, *Love at Goon Park* is a marvelous read. While I'm still not wild about Harry, I now understand a little of what drove this compulsive, dedicated man." —*New Scientist*

"It will make you think." —*The Seattle Times*

"Does a vivid job of unraveling Harlow's eccentric personality . . . makes a powerful case for his influence on everything from the treatment of premature infants and abused children to our most basic contemporary ideas of child rearing. Blum also defends him against his many detractors, including '60s feminists, who saw Harlow's emphasis on the maternal role in child rearing as a new attempt at domestic enslavement, and animal activists, who haunted him late in his career. Paradoxically, as Blum notes, Harlow partly laid the groundwork for those very activists with his studies of primate intelligence. Ultimately, *Love at Goon Park* does what the best works of biography and cultural history do (and this book is a mixture of both), presenting the evidence in as complex and nuanced a form as possible, and leaving the reader to make the final judgment." —*The Village Voice*

"Rivetingly recounts Harlow's work while examining the man himself . . . excellent." —*Library Journal* (starred review)

"Moving." —*Austin American-Statesman*

"A wonderfully written and maddening book, provoking, by turns, both delight and horror . . . Blum's greatest feat—more so than having written the type of cultural history that tingles with the discovery of new ideas—is that you neither worship nor revile Harry Harlow by the end of *Love at Goon Park*. You are humbled by his brilliant work, torn apart over his cruel methods, and ultimately grateful to live, and love, in a post-Harlow age."

—Salon.com

continued . . .

"Absorbing." —*Minneapolis Star Tribune*

"A biography of an innovative, controversial psychologist . . . A sympathetic and evenhanded treatment of Harlow's life and work—and an absorbing look at nineteenth- and twentieth-century notions of child psychology."

—*Kirkus Reviews*

"Surprisingly compelling . . . dwells carefully and passionately on the need for affection." —*Publishers Weekly* (starred review)

"Incredible as it may seem, half a century ago leading psychologists scoffed at the notion that affection was vital to an infant's flourishing. Deborah Blum brilliantly recalls this chilling era, and the scientist whose controversial experiments reaffirmed love's importance . . . science history at its best." —John Horgan, author of *The End of Science*

"Harry Harlow, whose name has become synonymous with cruel monkey experiments, actually helped put an end to cruel child-rearing practices. How these practices could ever have been advocated is only part of the puzzle presented in this lively biography. Blum does not shy away from the ethical questions raised by Harlow's research, yet reminds us that he was a complex man who won his battle with the scientific establishment so resoundingly that the outcome is now taken for granted."

—Frans de Waal, author of *The Ape and the Sushi Master*

"*Love at Goon Park* is the important story of the human need for love. Deborah Blum tells the engaging tale of Harry Harlow and his groundbreaking research with monkeys that proved our essential drive for social attachment. This book is not just good science writing, it's a great story."

—Meredith F. Small, author of *Our Babies, Ourselves* and *Kids*

"For generations of psychology students, the image of a baby monkey being comforted by a cloth doll is one of their most indelible memories of the subject. Yet even most psychologists know little about the brilliant, funny, and infuriating man behind the experiments. Nor do many people know about its context—the fall and rise of the concept of love in social science. Deborah Blum combines these elements into a gripping biography, written with intelligence, warmth, and panache." —Steven Pinker, author of *The Language Instinct*, *How the Mind Works*, and *The Blank Slate*

Love at Goon Park

Harry Harlow and the Science of Affection

DEBORAH BLUM

BERKLEY BOOKS, NEW YORK

A Berkley Book
Published by The Berkley Publishing Group
A division of Penguin Group (USA) Inc.
375 Hudson Street
New York, New York 10014

Cover design by Steven Ferlauto.

PRINTING HISTORY
Perseus Publishing hardcover edition / October 2002
Berkley trade paperback edition / February 2004

Library of Congress Cataloging-in-Publication Data

Blum, Deborah
Love at Goon Park / Deborah Blum.
p. cm.
Originally published: Cambridge, MA : Perseus Pub., 2002.
ISBN 0-425-19405-1 (pbk.)
1. Harlow, Harry Frederick, 1905– I. Title.

BF109.H346B58 2004
150'.92—dc22

2003060509

PRINTED IN THE UNITED STATES OF AMERICA

10 9 8 7 6 5 4 3 2 1

To Ann and Murray Blum

Absolutely my favorite parents

Contents

Preface

IN THE EARLY SPRING LIGHT, the Henry Vilas Zoological Park remains almost colorless, a place of black branch and pale ground. The upper Midwest shakes off winter slowly. Even as April dawns, the trees just hint at buds at the branch tips. The grass is beige stubble, the green bleached away by frost and snow. The faint bite of cold stirs the animals into protesting motion: the two lions prowling their woody enclosure, the one grizzly pacing his rocky ledge. This is a small zoo, after all, and a visitor walking fast can travel from reptile house to primate house—pretty much the length of the zoo—in a bare five minutes.

I'm a primate junkie myself. My footsteps, hollow-sounding on the cold ground, inevitably carry me to this building. I will stand admiring the plumy-tailed colobus—almost weightless in their grace—until my children drag me away. I have been known to loudly correct other visitors who tell their children that chimpanzees are monkeys. And I am eternally fascinated by the orangutans, with their ancient faces, gray as prehistoric stone, and their powerful bodies, deep bronze hair over fluid muscle.

Harry Harlow's research began here, in this pocket-sized zoo in Madison, Wisconsin, with a pair of orangutans. Those old apes are long gone, as are the old primate cages—smaller and sparer than the big bright enclosures of today—and Harry himself died some twenty years ago. I come here anyway, some days, as if I could find his ghost still watching the deliberate movements of the orangutans, just feeling that first tingle of kinship, the recognition that the animal on the other side of the bars is also watching you.

Until now, I had not realized how much writing a biography is an invitation to a haunting. For more than three years, Harry Harlow has inhabited my life. He's been an uncomfortable resident, as prickly as a spectral hedgehog. When I first started the book, my editor sent me an essay on whether a biographer must like her subject. It was an apt question. I didn't know whether I was going to like Harry Harlow. I thought, correctly, that he would be a sharp-edged subject—fascinating and troubling, and underneath the prickles the velvety gleam of brilliance. I wanted to write about him for exactly that uneasy mixture. Simple subjects, sometimes to my sorrow, have never interested me. I wanted to see his ghost walk, I suppose. I wanted to remind people that his work stays with us, still offering insight and promise. I thought—I still think—that he's been forgotten too soon.

You would have to call me an unlikely chronicler of Harry Harlow. Certainly, many of his friends have thought of me that way. I first wrote about him, more briefly, in an earlier book, *The Monkey Wars*, which explored ethical dilemmas of primate research. I wrote that book almost ten years ago—as I said, I'm a primate junkie. When I was done, my editor at Oxford said to me, "Harry Harlow is the most interesting person in this book. Would you be interested in writing a biography?" I could hardly refuse fast enough. I wanted a change. I was packing my monkey bags away and starting on a book about sex differences. Beyond that, my Harry Harlow chapter had made a lot of people angry. That vision—apples to oranges with this book—looked at Harry through the lens of the animal rights movement, which loathed him. Not surprisingly, this was not a focus that appealed to Harlow supporters. Many of them didn't want to talk to me again. Ever. Back then, I didn't want to talk to them, either.

So I started researching *Sex on the Brain* and found myself, in the course of exploring biology of behavior, talking to all kinds of primate researchers. That still didn't make me reconsider. It was thinking about children and parents who love them—and parents who don't—that really brought me back to this chilly path at the zoo. A couple years after the second book, when I was still writing about biology and behavior, I contracted with *Mother Jones* to do a two-part series on the science of neglected children.

It is so hard to do this kind of reporting, to deliberately immerse oneself in the bewilderment and grief of children pushed away by their parents, that I have to take a minute to say here how much I admire the counselors and child advocates who stay there, who live there the way journalists do not, who work to salvage lives. They are genuinely unsung heroes. While I was exploring the power of those parent-child relationships, I started thinking about other unrecognized heroes. "This is Harry Harlow's work," I realized. I rethought what he had done, not the primate research so much but the pure power of it, the way that it forced you to confront how much relationships matter in life.

And that's this book, partly a biography of Harry Harlow, partly the biography of a surprisingly recent idea in science—that love counts. A book is always a journey and at the end of this one, I asked myself whether I had learned to like Harry Harlow. Many of his family members, friends, and colleagues did eventually agree to talk with me. "I didn't like your first book and I don't really like you," one scientist told me. "But I want to have input." I'm not the only one to wonder whether I would develop affection for Harry. Easy question, tricky answer. He makes me laugh, even secondhand. He makes me think about friendship and parenthood and partnership in ways that I never had before. He still seems to me an edgy companion. And he seems wholly real. So, like Harry, the answer is complicated. Sometimes I do like him, sometimes not at all. In the end—it's the both that makes him such a terrific subject for a biography—exasperating, sometimes, enchanting other times, never boring. And his weaknesses give a curious strength to his work—he was bitterly honest, sometimes to his own detriment. He was willing to take his personal problems—loneliness and isolation and depression, even—and use them in his research.

I do return to the ethics of primate research in this book, but only briefly. This is a biographical story and during Harry Harlow's lifetime animal ethics did not dominate the discussion. So you will find those questions in the final chapter. I have touched lightly on other aspects of Harry Harlow's research, as well. He was such an eccentric character, he had such a restless intellect, that if I had followed every idea he pursued, I could have meandered through half the history of psychology. So I didn't write about his interest in neurochemistry or his experiments with infant retardation or his early work on brain structures. I didn't include every funny story or interesting anecdote that people shared with me, either. Oh, I wanted to put in everything. I would have crammed in every recollection if I could only have had another five hundred pages or so.

And I would—primate junkie again—have written more about his research into the natural intelligence of primates if I could have persuaded myself that monkey cognition was central to the story of love. So I acknowledge in advance that this is not an all-encompassing biography, or a detailed history of psychology. It is rather a journey with one scientist, a pursuit of the role of relationships in human behavior. Everyone needs "a solid foundation of affection," Harry once said, and this book is about his efforts to dig down to that emotional bedrock.

I wonder about those fundamental lessons as I stand inside the primate house, shutting the glass doors on the slow thaw outside. I watch the orangutans with their Stone Age faces and think about how we learn about love. The orangutans at the Vilas zoo have a new baby. The mother holds it, heart to heart, as if letting go would violate all the natural laws of life. Perhaps science is finally catching up with common sense, as Harry liked to say. Perhaps the answer is as simple as the view through the glass: mother and child so close together that you might imagine the two hearts beating as one.

Deborah Blum
Madison, Wisconsin
April 2002

Acknowledgments

THEY SAY THAT WRITING A BOOK is a solitary operation, but they lie. Oh, there are moments, when one is shackled to the keyboard, that it feels like a confinement. Moments stuck in the quicksand of some sentence, when you realize that no one else can save you from being sucked down. There are maybe too many moments when the book sounds so loudly in one's thoughts that it's almost difficult to hear anyone else over the thunder of the language inside.

But mostly, writing nonfiction is a community project. No one tells a story such as this, the story of another's life, the story of a changing science, without help, and lots of it. These travels through the life and times of Harry Frederick Harlow could not have been accomplished without many guides. The words that come to my mind are generous and patient and those words describe the people who helped me assemble, fragment by piece, the intricate mosaic that makes up a life.

I owe endless gratitude to Robert Israel, Harry's oldest son, and Helen LeRoy, one of Harry's closest colleagues and friends, for their time and many kindnesses. Bob Israel, who lives on one of the loveliest hillsides in the Pacific Northwest, invited me to his home and spent hours talking to me about family history, lending me photos, copies of his father's drawings and poems, and even sending me pages from an unpublished autobiography that he found boxed up deep in a closet. Helen LeRoy, here in Madison, who has carefully preserved the letters and documents and artifacts that illuminate Harry Harlow's life, put up with a near constant barrage of e-mails and questions and requests for documents and meetings to go over facts; and she remained ever gracious and, equally important, meticulous about the accuracy of the story.

I also want to thank Harry's son, Rick Potter, Harry's daughter, Pamela Harlow, and his brother-in-law, Robert Kuenne, for their help with questions about family life. My gratitude also to the gracious people in Fairfield, Iowa, Harry Harlow's home town, who went into the local archives for me, located old neighbors, and drove me all through town so that I could understand the social geography of Jefferson County. In particular, I would like to thank Jim Rubis, Ron Gobble, and Hazel Montgomery.

I wish I could list every one of Harry's former students and colleagues who dug into their own files to find correspondence and dug themselves out of my avalanche of questions with kindness and humor. I particularly want to mention Robert Zimmermann, who worked with Harry on the first cloth-mother studies. Bob and his wife, Marian, invited me to their Michigan home, lent me copies of photos—including some of the old glass slides from the 1950s—and answered every follow-up question with great thoughtfulness. William Mason, of the University of California in Davis, a psychologist whom I have known and respected for years, and one of Harry's most influential postdoctoral students, talked me through not only the past but many of the present nuances of the science. Duane Rumbaugh, at Georgia State University, is a hero of this book. He sent me copies of illuminating life-long correspondence with Harry and guided me through much of the history of early animal intelligence research, always with his particular gift for emphasizing the value of the animals as well as the science. Jim King at the University of Arizona was everything a book researcher could hope for. He and his wife, Penny, played host, took me to Harry's former home there, arranged a dinner with the smart and funny former head of the Arizona psychology department, Neil Bartlett, and his equally smart and funny wife, Olive, and got me thoroughly drunk on one memorable evening that began with Rob Roys and ended in a dream-like haze of after-dinner drinks, whatever they were, I'm not quite sure.

Steve Suomi, head of the National Institutes of Health (NIH) lab of comparative ethology invited me to his laboratory to talk—despite his doubts about my trustworthiness—and followed up for months responding to a thousand nit-picky questions. Jim Sackett met with me in Seattle to provide a candid and courageous interview, especially considering that he continues to be targeted by the local animal rights community. Seymour "Gig" Levine met with me both in San Francisco and in Madison, and as always charmed me into paying for lunch, gave me a hard time, and patiently explained both his work and its context. He has threatened to quit speaking to me for ten years now and I appreciate that he continues to answer my questions anyway.

I would like to also thank Dorothy Eichorn, who met me on a spring Sunday on the grounds of the old state mental hospital in Napa to talk about her long friendship with Harry Harlow; Albert Hastorf, keeper of the Lewis Terman gifted files at Stanford, who invited me into his office on a bright weekend day, handed me the files on Harry's first wife, Clara, and left to play tennis, saying "lock up when you leave"—which I took as an enormous compliment; the wonderful staff of the Archives of the History of American Psychology in Akron, Ohio, in particular, David Baker and Dorothy Gruich; and Margaret Kimball and Henry Lowood of the Stanford archives, who walked me through the documents I needed from Harry's early history there. And I am indebted to Robert Hinde, Leonard Rosenblum, Melinda Novak, Judith Schrier, Ed Tronick, Meredith Small, Stephen Bernstein, Sally Mendoza, Kim Wallen, William Verplanck, Irving Bernstein, Larry Jacobsen, and Richard Dukelow for their generous help and invaluable perspective. Two people interviewed for this book died before it was finished and I would like especially to mention their patience and humor with the process. They are Art Schmidt and Richard Wolf, and they are both missed.

The University of Wisconsin at Madison supported this book from the beginning, and in particular, the Graduate School provided funding for student researchers and summer salary so that I would have extra time to work on the manuscript. I was blessed with some dedicated and extremely smart graduate students who read old psychology texts, tracked down long-lost contacts of Harry Harlow, and made a real contribution to the depth of this book. They are a group of outstanding young journalists named Tina Ross, Brennan Nardi, Suzanne McConnell, Krishna Ramanujan, Morgan Hewitt, and Maggie Miller.

I owe an infinite debt of gratitude to Robin Marantz Henig, Kim Fowler, and Peter Haugen, who read the book in its earliest, most chaotic version and helped turn it into an actual story. George Johnson and Shannon Brownlee were enormously helpful with some of the trickiest chapters in the book.

I am blessed, as always, in my agent, Suzanne Gluck, who believed in the Harlow story from the start and helped me see its potential as well. And I am doubly blessed in having the best possible editor in Amanda Cook. Amanda is such a good editor, so smart and so supportive, and so gifted in her ability to improve a story, that I have been congratulating myself on my good fortune almost since I started the book. Perseus's talented and meticulous production staff improved and clarified the story I wanted to tell and then packaged it beautifully and I would particularly like to thank copy editor Jennifer Blakebrough-Raeburn, who has a wonderful literary mind, and senior project editor Marietta Urban. And I also had the good fortune of being teamed with Perseus's terrifically smart publicist Lisa Warren.

Last, but not least, as they say: This book is about family and love and partnership and relationships and there is no way that it would have been finished without the love and support of mine. I've been wrestling with the Harry story for so long now that both my children, Marcus and Lucas, also call him by his first name. So does my husband, Peter Haugen, who has held down the home front too many times while Harry, my laptop, and I were hunkered down in my basement office together. The three of them kept me from disappearing entirely into the book, partly because they never fail to remind me that nothing matters more than those we love.

PROLOGUE

Love, Airborne

In a white room, two men are talking about love. One of them stands keenly upright, pressed into a deftly cut suit. The other is less elegant: slight, dark-haired, a little stoop-shouldered, shrugged into a floppy lab coat. Both their voices sound hollow in the pale space around them. The room seems glossy with cold. Nearby counters are polished to an icy sterility. Metal and glass equipment gleam bluish in the wash of fluorescent lights. Against this background—chilled essence of laboratory—the speakers sound like men out of place and time, their conversation absurdly soft with talk of poets and love songs, starry nights, and daytime dreams.

Or perhaps they are just ahead of their time. At this moment, in the close of the 1950s, no one stands in a laboratory to discuss love in these terms. Even psychologists—those perpetual students of human behavior—aren't lobbying to include warmhearted affection among the charts and the graphs and the calibrated machinery. Experimental psychologists have been rejecting the notion of love as good research material for years. Powerful psychologists have made it clear that fuzzy and sentimental emotions are the stuff of fiction, not of research reports. Researchers who study human relationships prefer to avoid using the L-word. You can still open the acclaimed history of *Psychology in America,* by Stanford's Ernest Hilgard, and find the word "love" missing entirely from the subject index.

So it's a professional gamble for the small man in the lab coat even to have this conversation. He *is* an experimental psychologist—a stubborn, scruffy, middle-aged researcher named Harry Frederick Harlow who happens to believe that his profession is wrong and doesn't mind saying so. Of course, he's often been told that the problem lies with him. The unexpectedly outspoken son of a poor family from Iowa, he's developed a habit of scrapping with mainstream psychology. Professor Harlow has already been asked to correct his language: He's been instructed on the correct term for a close relationship. Why can't he just say "proximity" like everyone else? Somehow the word "love" just keeps springing to his lips when he talks about parents and children, friends and partners. He's been known to lose his temper when discussing it. "Perhaps all you've known in life is proximity," he once snapped at a visitor to his lab at the University of Wisconsin in Madison. "I thank God I've known more."

How close do you have to be standing to connect with a person? Harry liked to ask that question, drawling it out with a nice sarcastic edge. Three inches? Four? Could you build a relationship at a distance of six inches? His colleagues, as they told him, saw no need for mockery. He could choose other scientific terms if he didn't like proximity. The scientific vocabulary also offered attachment, conditioned response, primary drive reduction, stimulus-response, secondary drive reduction, object relationship—the last if you wanted to be Freudian about it. Why bring love into it?

And now here's Harry Harlow, on national television of all the damned places, with his intimate vocabulary and his insistence on emotional relationships. The conversation in the laboratory appears on a CBS show called *The Measure of Love.* The program is the 1959 premier of *Conquest,* the network's Sunday evening science show. In the entire half-hour presentation, the word "proximity" never crosses Harry's lips.

Charles Collingwood, a respected CBS journalist, is the man in the elegant suit. On camera, Collingwood stands authoritatively tall. Harry Harlow looks small by comparison, dwarfed inside the ubiquitous lab coat. He has a square face, dark eyes under near straight brows, short dark hair slicked determinedly back. His voice is a little high-pitched, smoother than Collingwood's rumble.

But the voice of science is unexpectedly the voice suitable to a pulpit, slightly singsong in its cadence. There's music in the way Harry assures us that it is possible to make real what had previously been "undefinable and unmeasurable." As he talks, one might even believe that love is substantial enough to be decanted into test tubes. When it comes to love, "your guess is as good as mine," Collingwood says to the audience, "but guesswork is not the way of science and *this,*" and his gray granite voice deepens a notch, "this is a scientific laboratory." At the start of the program, Collingwood stands holding a monkey in one hand. The monkey is a bright-eyed baby, a natural mohawk of fluff crowning its head. It nestles in Collingwood's curved hand like an egg in a cup, tiny fingers curled over the edge of his palm. Harry Harlow, after all, is a primate researcher, a pioneer in the emerging science of understanding monkey behavior as a way to understand us. Collingwood gestures slightly to emphasize that point, the monkey riding the sweep of motion: "In this laboratory, there are approximately 120 rhesus monkeys; the subject of a study that wants to know the answer to the question: What is an infant's love for its mother?"

There's little trace, here on *Conquest,* of what some would say is the off-camera Harry Harlow, none of his well-known irreverence. This is a man who when a graduate student points out a golden and luminous moon snaps: "Been there a long time. I've seen it before." None of that wisecracking irritation shows now. This shiny faced, sweet-talking preacher of a scientist seems wholly absorbed by the beauty of the subject. The man on camera reveals little of the man who lives at the lab, dawn to dark, fueled by coffee, cigarettes, alcohol, and obsession. Okay, maybe the obsession slices through. He's completely in the argument, trying to convince the world that if science will just pay attention, we could learn the measure of love, cup it in our hands, almost as Collingwood cradles the little monkey.

"Now, Mr. Collingwood, wouldn't you say that if you frightened a baby, that if it went running to its mother, was comforted, and then all the fear disappeared and was replaced by a complete sense of security, that baby loved his mother?" he asks in that coaxing voice.

"Sure," Collingwood replies, casually. Sure, of course. Who wouldn't believe that love was, at its best, a safe harbor—a parent's arm scooping up a frightened child, holding it heart to heart? It's hard to believe, in retrospect, how many powerful scientists opposed this idea. "In psychology, love was smoke, mirrors, bullshit, and that was exactly what everyone was telling Harry," one of Harry's graduate students recalls. It took courage, probably more than anyone at CBS appreciated, to look straight into the camera and contradict the professional standards of the time.

There's a moment on the *Conquest* show when one of the Wisconsin experiments is displayed. It creates exactly the sequence that Harry described. The scientists send out a mechanical monster, maybe eight inches tall, that resembles a cross between a space alien and a dragon with its flashing eyes and black bat wings. "It looks diabolical," says Collingwood. "That's just the way a baby monkey feels about it," replies Harry—and almost as he speaks, the baby monkeys take one look at this terror and go airborne.

They fly like guided missiles—a perfect arc of child to mother. Look, says Harry, mouth curving. One of the baby monkeys, now firmly lodged against mother, is screeching angrily at the monster, threatening it: Back off, you. I'm with my mother now. If a measure of love is the way we shelter each other, you can mark it clearly in that fluid and beautiful flight line to home.

The two men watch silently. Harry doesn't have to add anything. He knows it, too. He can step back and let the relationship reveal itself. Baby to mother, arrow into the heart. He does have a take-home message though, as he stands here in his baggy coat and talks up the importance of simple affection. The message has enough potency that you can understand why it might be worth contradicting more than fifty years of scientific dogma.

In this conversation about love, the two men have different goals. Charles Collingwood has come to Madison, Wisconsin, to illuminate an unusual experiment and to make some good television. Harry Harlow is there to help him. But he's also trying to foment a small revolution, taking the chance to provoke the argument even during this flickering black-and-white moment on Sunday television.

We begin our lives with love, Harry says, looking directly at the camera; we learn human connection at home. It is the foundation upon which we build our lives—or it should be—and if the monkey or the human doesn't learn love in infancy, he or she "may never learn to love at all." He looks absolutely confident in what he's saying—as if there were no furious ongoing debate, as if he spoke for his profession. Arguing his point as an outsider is a skill that Harry Frederick Harlow has honed since childhood. He's more than willing to stand on behalf of that improbable, unreliable, elusive emotion called love, to gaze into the camera lens and say: Listen to me. I've got something that you need to hear.

ONE

The Invention of Harry Harlow

Parental love, which is so touching and at bottom so childish, is nothing but parental narcissism born again and, transformed though it be into object-love, it reveals its former character infallibly.

Sigmund Freud, 1914

HE WAS BORN OUT OF place, a dreamer and a poet planted in the practical Iowa earth. As unlikely as a rose in a cornfield. The childhood of Harry Frederick Israel—he would become Harry Harlow, but that's a later part of the story—often made him laugh in retrospect. He was such a funny little misfit of a child, hemmed in by the orderly fields, too often dreaming down those rows of green and gold to the point where they met the rim of the sky.

This was southeastern Iowa, after all. Everyone grew up amid the cornfields. At the dawn of the twentieth century, the landscape was a study in domestication. Paradoxically, that very neatness made Iowa a revolutionary corner of the country. Not even a hundred years before, the land had belonged to lynx and wolf, deer and buffalo, the elusive catamount, and the bright copper fox. Tall-grass prairies and wooded hills, undisciplined rivers that had never seen a levee, forests with familiar trees such as maple and birch and forgotten ones such as linn and ironwood. The Fox and the Sac tribes once hunted here, gathered wild plants, quarreled over territorial boundaries, called it home.

The old settlers—Iowans think the term "pioneer" sounds too transient—began transforming the land in the early nineteenth century. The little town of Fairfield, where Harry was born many years later, was chartered in 1836, neatly laid out around a traditional town square. For decades, it retained a frontier quality. Until the 1870s, hogs were allowed to run through the square. When the mayor finally insisted that livestock be penned, pig owners angrily protested this affront to liberty. People paid their bills with what they could grow or raise. The town doctors accepted everything from chickens to tomatoes. The pharmacies on the square sold Indian remedies to their customers, tidily packed cloth bags with chamomile flowers for measles and slippery-elm bark for pneumonia.

Science was something distant, not quite real and not all that important. "Few knew or cared that the world was filled with innumerable fascinating creatures or that the history of the earth was written in the rocks beneath their feet," wrote the Fairfield historian Susan Fulton Welty in a loving tale of her hometown. In the late nineteenth century, some Fairfield high school students formed a science club. They were enthusiastic, but they found the subject mysterious at best. One of the first meetings raised the question "Is a Bat a Bird?" The members were mostly nature collectors. They packed their clubhouse with pinned insects, dried flowers, the brittle remains of ferns and mosses, and assorted bones. At one point, club members assembled almost the entire skeleton of a horse, built from bleached bones found tumbled in a nearby pasture.

By the time Harry Israel was born, the frontier had been tidied away. The town square was neatly paved. The Sac and the Fox had mostly vanished, pushed to the west. The herbal remedies had been replaced by a red-brick hospital and more European-style medicine. The woodlands and feathery fields were plowed, tilled, and rotated into submission. Even the science enthusiasts had given up bone hunting. The local high school now taught the study of nature, "with especial attention to the highest of vertebrates, Man himself." Harry would have preferred it just a little less, well, predictable. Years later, he would confess that completely orderly science bored him. He could never quite accept rules as absolute. He was never really convinced that "Man himself" was an example of evolutionary perfection. A work in progress, maybe. He would have been happy to argue the point—if it had been open for debate in Fairfield. His family would have said that Harry was born to argue. So would his peers. When he graduated from high school, this quote appeared under his yearbook picture: "Though rather small, we know most well, in argument, he doth excel."

He was born on a Halloween evening, October 31, 1905, at his family home in Fairfield. "Within thirty minutes I had precipitated a violent family quarrel," Harry once wrote. His Aunt Nell had come all the way from Portland, Oregon, and wanted to hold the baby first. But his two older brothers begged her to take them on a quick trick-or-treat outing. When the three of them returned, baby Harry was lying cozily in his Aunt Harriet's lap. "This was a situation in which better late than never did not pertain," Harry would joke later. Harriet lived just around the corner in Fairfield. Nell had traveled hundreds of miles. And the ungrateful baby's parents had named the child *Harry*. In family lore, the story of his birth always resounded with the ensuing thunder.

"Another memory which I do not have happened when I was three," Harry wrote years later in an unpublished memoir. The entry was typical of the way he recounted his childhood—always flippant about growing up in Iowa. As he told the story, when he was a little boy, he owned a porcelain child's potty, which he loved. He would carry it around the house with him. One day, according to his mother, "guided by uncontrolled scientific curiosity, I dropped a large stone on the potty's bottom to see what would happen." He sobbed over the pieces for days afterward. An incurable punster for most of his life, Harry wrote that his grief was probably caused by his having hit "rock bottom."

His parents were Alonzo Harlow Israel and Mable Rock Israel. If Harry was something of a misfit, that standard was perhaps first set by his father. Lon Israel—he loathed the name "Alonzo" and as an adult refused to respond to anyone who called him that—had hoped to be a doctor. He gave that up, though, dropping out of medical school in his third year to marry Mable Rock. Lon never quite found anything else that he liked as much as the study of medicine. He reluctantly tried and happily abandoned farming. He tinkered with what Harry called "intermittent, unsuccessful inventing." Lon experimented with home appliances, and once even developed a small washing machine. He dabbled at running a garage and battery business, teaching himself about mechanics by reading books and manuals in a weekend frenzy. He started a small real estate business with his father. Eventually, Lon and Mable bought a general store in a small town near Fairfield and settled there. Harry's parents had been married for ten years and were in their mid-thirties when he was born. At the Fairfield public library today, there is an archived photo of Lon on his wedding day: a slim man with a pointed chin, dark eyes under deep brows, a thin mouth just tilted into a smile at the corners. There is also a photo of Mable wearing a lacy white dress that seems to float at the edges. Mable was barely five feet tall. In the picture, she is as delicate as a fairy, fine-boned and graceful in her posture, her shining dark hair pulled smoothly back from a small, rather beautiful face. The Israels had four sons, in this order: Robert, Delmer, Harry, and Hugh. The boys all had their mother's slight build, their father's brown eyes and heavy eyebrows. In Harry's face, one can also see Mable's finely drawn features and slightly squared, stubborn chin.

Harry remembered his parents as being determined that their children would grow beyond them. They had to fight for that—another lesson learned early. He was just three years old when his older brother Delmer was diagnosed with Pott's disease, sometimes called tuberculosis of the spine. Lon Israel had outguessed the local doctor on the ailment. Disturbed by the increasingly warped look of his son's back, Lon bent an iron rod into the same odd curve. He sent the bar to a research hospital in Chicago, where doctors made the diagnosis from the distinctive bend in the metal. They recommended that the boy go to a warmer, drier climate—then the standard remedy for TB. Frightened for their son, the Israels sold their house and moved the family to New Mexico. Short on money, they camped in a small canyon outside Los Cruces. Delmer's health did improve in the brilliantly lit New Mexico air. But the family, already poor, grew more so. They lost their remaining possessions in a season of wild spring flooding. At one point, Lon Israel was forced to carry his children out of a rising stream when it flooded through their tent. In little more than a year, the family returned, near destitute, to start over again in Fairfield.

His parents, Harry said, "literally lived for their children. Fortunately, they did not have enough money to be really indulgent." Not that he wouldn't have enjoyed a little more indulgence—or extra affection. His own research would lead him to realize, many years later, how much he had felt like an afterthought and how much he had minded. "I remember my mother as a tiny, beautiful, hardworking, and efficient woman who reared four sons, and probably a husband, ably, lovingly, providently. I always thought of her as a person who loved me dearly, and I am sure she did." With Delmer's illness, though, he suspected "she was probably hard pressed to shower affection on others." Harry was just a toddler when his brother fell ill. His mother was there, near the home, physically—just not quite all there emotionally for a small, shy younger brother. "I have no memory of partial maternal separation, but I may have lost some percentage time of maternal affection, and this deprivation may have resulted in consuming adolescent and adult loneliness."

Almost five thousand settlers now occupied Fairfield. Ornate buildings, topped with towers and ramparts, housed shoemakers, grocers, barrel makers, tailors, druggists, clothing stores, furniture stores. The square was a gathering place for the farmers who now ploughed the surrounding country. Even in winter, when the farms were iced over and Fairfield's streets were deep with snow, farmers came to town. They simply took the wheels off their wagons and replaced them with heavy, ironclad sled runners. Fairfield's children used to play street games in which they jumped from farm bobsled to farm bobsled. They called the game "hopping bobs," and, as one sled hopper recalled, the farmers were cheerfully tolerant of the leaping children.

Farming was the breath of the town. Harry's father had himself listed farm properties during his real estate venture years. The local high school balanced traditional academics and agricultural education. Girls were required to take domestic art and science, courses such as "How to Cook to Please the Men." The comparable track for boys was farm management, from crop rotation to pest control. The wood-frame homes, brick businesses, and orderly streets of Fairfield merged almost seamlessly with the outlying farms and orderly fields that surrounded it. And here was this quiet dreamer of a child, without a shimmer of interest or ability in even managing a garden. Many years later, Harry's oldest son, Robert, would recall that the few times his father attempted yard work, he routinely uprooted prized bedding plants. "It was always, 'Call the yard man' at our house." Harry had no interest in geraniums and nasturtiums as an adult, and less in tilled fields as a child. He liked to write poetry and draw pictures. He recalled once completing an essay assignment that "didn't sound right" and deciding not to hand it in. Later he realized that he had spontaneously written the essay in blank verse. It wasn't just that he could write verse—an impractical talent if there ever was one—he actually liked it. One of his favorite assignments came in the eighth grade. He and his fellow students were told to compose a four-line verse on the "benefits and beauties" of daily tooth brushing:

> Students filtered into the class expressing hate and hopelessness at the assignment. I rose to the rescue. By ten minutes of nine, I had completed fourteen verses for fourteen students—aside from the best, which I kept for myself. The teacher was pleasantly surprised at the literary level of the class and she selected five for indulgent praise. All five selected were mine but the one I selected for myself was not among them. It dawned on me that I was a better author than critic.

Mostly he was bored. "My high school academic career was not totally distinguished. I ranked thirteenth out of a class of seventy-one whose average IQ was below 100." The top twelve, he noted, were all girls. He did outscore the entire senior class on an aptitude test created by the University of Iowa. The results were put on a big blackboard—in those times, educators didn't consider sparing the feelings of the students. "I was about two standard deviations ahead of my nearest competitor, who was the female class valedictorian and the girl whom my grandfather hoped I would marry because she was the only daughter and granddaughter of a wealthy family," Harry wrote in his memoir. Not in this lifetime was Harry Israel going to marry into a commitment to stay in Fairfield. He planned to be somewhere else—someone else. In the 1923 yearbook, the year of his graduation, his senior class photo shows an unsmiling boy. He has downcast eyes, a shadow of long lashes about them, smooth dark hair, lips slightly turned down at the corners. In the same yearbook, students are asked to say what they wish to be when they grow up. The dreams are mostly small ones, happy ones. One wants to be a teacher, others want to be pretty, lovable, a farmer, a musician, a farmer, a singer, a farmer. Harry Israel's wish? At the age of seventeen, he wanted to "be famous." He made a prediction, though, for his more probable outcome: He would simply end up insane.

The Israels, you might say, were not a routine Fairfield family. Most of the local businessmen were not building experimental washing machines in their garages. And almost all the townsfolk met and gathered and socialized at one church or another. Fairfield and the surrounding Jefferson County were a paradise of churches at the time. The Israels' home sat in the gothic shadow of the First Methodist Church, a looming brick structure just around the corner. In the county's first hundred years, eighty-five churches were built: twenty Methodist, nine Baptist, seven Lutheran, six Presbyterian, four Catholic; and Dutch Reformed, Christian Science, Adventist, more. One of the few failed congregations was the Episcopalian, which had been the Israels' chosen house of worship. When the modest building burned down, though, the small congregation drifted into other houses of worship.

Lon absolutely refused to drift. Even in New Mexico he had attended the Episcopal church, wearing his faded jeans and battered hat but going every Sunday. It was Lon's church or no church and, after a while, the family simply stayed home. The boys played cards on Sunday mornings and watched their neighbors, dressed in their best clothes, walking to church. It created something of a distance. People made friends at church, traded recipes and gossip, and planned dinners together. In a Christian town such as Fairfield, your neighbors noticed when you didn't take your place in a pew.

"It was a small town," says neighbor Hazel Turner Montgomery, now ninety-seven, who once lived around the corner from the Israels. She remembers as a child visiting the family, walking over to read to Delmer. She would sit in the parlor, while he was strapped to a backboard, and while away the slow afternoons. Montgomery is a small, bright-eyed, friendly woman with a fluff of silver hair. She has always enjoyed the company of others, but she's not so sure the Israels felt the same way: "You didn't see them walking out. Everyone knew they didn't go to church. I don't believe they were, well, a very sociable family." And she's wondered sometimes whether they were lonely. She remembers that Delmer never was quite ready for her to leave.

Lon and Mable Israel didn't raise their children to be joiners or conformers. They wanted them first to think for themselves, and fitting in took second place. If they were distant from Fairfield in some ways, lonely as children, that made them tighter as a family. They competed for honors in school, sharing achievements during meals. By family accounts, Delmer was the brightest and Harry the most competitive. He would sometimes try to catch his brother out: "Who won the battle of 1066?" he would demand, right in the middle of breakfast.

The Israels wrote and staged backyard plays. The parents insisted that all their sons study music. Robert was a genuinely talented musician, Delmer a reasonable saxophone player, Hugh a credible piano player, and Harry a dogged one. Harry played the piano, but he was never very good at it, at least compared to the artistry of his eldest brother. "My talent lay at the exact opposite end of the scale," Harry liked to explain.

He loved art, though, and throughout his life made time to draw, creating beautiful, fantastical landscapes with ink and colored paper. Even after they entered college, Harry, Robert, and Delmer dreamed together; they invented a fantasy country, The Land of Khazoo, into which only clan members were welcome. "In the Khazooan ranks, you'll find a few friends, well good, nothing could be more valuable—but the inner clique you'll find to be a family affair and totally understood so far by Del, Harry and I," wrote Robert in an explanatory letter to their father. The artist among them, Harry, was designing a shield that would bear their motto. It had three words in the crest: "Israel über alle."

There was never a doubt that the Israel boys would go to college. Their parents saved for it, their grandparents chipped in, even their aunts contributed. "Our parents were determined," Harry said simply. If he yearned for a life beyond Fairfield, his parents wanted it for him and his brothers as well. Harry, Delmer, who gradually eased into good health, and Hugh all went to Stanford University. Harry could hardly wait to go. He was almost there, in the different country he'd dreamed about, there where the Iowa cornfields brushed the horizon.

He was the only one though, of the California-bound Israel boys, who saw that promise and he was the only one who stayed to finish his degree at Stanford. Delmer dropped out of his law program, got married, and ran a sports equipment store in Palo Alto. At least a few people still remember him for his razor-sharp tennis game and for his shop's beautifully re-strung racquets. Hugh studied oceanography. Unlike his brothers, though, Hugh was an Iowa boy by nature and the cold glitter of the Pacific only made him miss the gentler landscape he knew best. He returned home without earning a degree.

Their parents had moved twenty miles from Fairfield to the even smaller town of Eldon, where they were running their general store.

Hugh joined them in the business and stayed there. Robert, the eldest, earned an M.D. in psychiatry at the University of Oregon in Eugene, and spent the majority of his career as chief psychiatrist at the state mental hospital in Warren, Pennsylvania. Harry—who stayed closest to Robert of all his brothers—once sent him a poem, called "The Madhouse at Midnight":

I'm in this institution
On the pretense that I'm insane
But this, as everybody knows,
Is nothing but a guise
The reason that I'm here
Is very easy to explain
The War Department Thinks
I am a pair of Russian spies
Sometime I'm going to leave this place
I haven't picked the day
I'll simply push the buildings down
And calmly walk away
I'll build a little railroad
That reaches to the moon
And run a little subway
From New York to Neptune
I know where all the money
In the universe is stored
I'm the nephew of Napoleon
And cousin of the Lord.

Harry had always understood escape dreams. Hadn't he been dreaming them himself for years? And at Stanford he found something like a railroad to the moon, a way to soar beyond the domesticated landscape of his childhood.

◦ ◦ ◦

To a boy raised in farm country, this young university was as improbably different as a lunar landscape. Even in the 1920s, coastal California was still part wild. Stanford itself, just south of San Francisco, was a small, civilized outpost in the windblown hills. To the west was the dark blue glimmer of the ocean and to the north the still darker rise of the Sierra del Monte Diablo.

Harry liked to tell the story of how he arrived, or almost didn't, at Stanford in 1924. He had spent a year at Reed College in Portland, Oregon—close to his Aunt Nell—when his parents decided he would be better off in California. Harry just wanted to study somewhere that made him think. He'd coasted through high school. He didn't want to sleepwalk through college, too. "The first course to stimulate me intellectually was a freshman course in zoology that I took at Reed College. But they made me dissect a dead frog and I *despised* dissecting dead frogs. So I decided to find a science that was like zoology but that didn't specialize in dead frogs."

Delmer had already been admitted to Stanford. Their parents telegraphed Stanford ten days before the fall semester started and asked whether both brothers could attend. As Harry recalled it, Stanford was unenthusiastic. But there was a mechanism to give last-minute applicants a chance. The university gave a special examination and would admit thirty additional students, those who scored the highest on the test. Harry was among about a hundred other last-minute prospects. He looked around the exam room and gained an impression of being surrounded by giants: "One glance at my fellow applicants convinced me that one half of the hundred were football players dredged up to round out next year's team." He found hope in that company, though. "I decided that if I could not win against this competition, I did not deserve to go to Stanford."

After clearing the admission hurdle, Harry enrolled as an English major. He still liked writing, and believed it was his strongest talent. His first semester thus was a terrible shock. He got a C+ in English. In a fury of disappointment, he switched his major to psychology, figuring that it would eliminate both the writing issues and the dead-frog problem. It's worth noting that both Harry and the instructor changed their minds. The same teacher later included him in the *Stanford Mosaic,* a collection of works by students who were gifted writers. And Harry began to accept that perhaps he wasn't an exceptional talent: "I rather believe that her first judgment was correct," he said after the *Mosaic* appeared. Still, he never lost the habit of playing with words. Even as a psychology major, he tried taking notes in verse, although he freely admitted that describing medical symptoms in the form of poetry didn't work that well:

Apathetic Annie was complacent and serene
Though suffering from paresis,
Consumption and gangrene
But Annie did not really care
Though life was nearly gone
For Annie had a tumor in the diencephalon.

Although he would share his rhymes freely later in life—leaving doggerel verses on his grad students' desks, mailing rhymes to friends and business associates alike—at Stanford, he kept the verse notes to himself. Apathetic Annie, Narcoleptic Nancy, and all their equally physically impaired friends were tucked neatly away. Somehow he was never quite sure that Stanford would really appreciate them. He might have been an exotic flower back in Fairfield, Iowa, and he might have been the most promising student in his high school class, but here at Leland Stanford's memorial university, bright students surrounded him. He felt dusted with corn pollen and self-doubt.

The central Stanford campus is a beautiful, arrogant place. Frederick Law Olmstead, designer of New York City's Central Park, laid out the university's landscaping. Boston-based architects were chosen to give the buildings an old Italian elegance. The resulting main quadrangle is brilliant with red tile roofs, fringed palm trees and, of course, the dancing, luminous light that refracts off the nearby Pacific Ocean. It illuminates the old campus. It washes over the Memorial

Church, over its Venetian glass murals and sternly carved Victorian moral sayings: "A noble ambition is among the most helpful influences of student life and the higher this ambition is, the better."

Harry tried to walk quietly around Stanford's elegant passageways and shining exhortations and, oh yes, the self-styled geniuses who ran the department of psychology. As a graduate student, he worked directly under Calvin Stone, an animal behaviorist and editor of the respected *Journal of Comparative and Physiological Psychology*, and Walter Miles, a vision expert who would eventually design night goggles for World War II fliers. Stone and Miles, in turn, worked under Lewis Terman, the flamboyant, red-haired, ultraconfident developer of the Stanford Binet IQ test who served as the department's chairman. Harry considered these three men—in both positive and negative ways—the fathers of his passion for the science as it existed and of his desire to change it.

He always called Walter Miles his moral mentor—although that was partly affection. Miles liked Harry, too. He went out of his way to give the young psychologist extra support. When Harry's money started running out—and he was reluctant to demand even more sacrifices from home—Miles gave Harry a job. Miles kept a colony of rats in the garage of his Palo Alto home. Harry would hurry over to his professor's home after a day of classes and help run the rats through experiments. He became friendly with the professor's family; although, as he noted, not too friendly: "From time to time Dr. Miles' disarmingly beautiful daughter dropped in and chatted on her way home from high school. Dr. Miles gently discouraged this platonic pursuit. He had higher aspirations for his daughter and so did she."

Harry was unoffended, and, frankly, uninterested. He was far more focused on making it through Stanford than on pursuing high school students. Although his brother Robert used to laugh about the girls that Harry had yearned over in Fairfield, at Stanford he didn't pursue any serious relationships. He was turning into Harry Harlow, beginning to develop the tunnel vision—not Israel über alle, but psychology before all—that would also characterize him through much of his life. And he learned, from the ways that Miles tried to help him, that colleagues could also be family.

Harry's major professor, Stone, was neither warm nor nurturing nor familial. But he was a scientist through and through. Stone approached his students almost as he did his experiments: with absolute insistence on getting it right. He was a dedicated believer in the animal model. Most of Stone's research was done in rabbits and rats. He studied the effects of brain damage on the sexual behavior of rabbits. He looked at the influence of diet on the sexual responses of albino rats. He explored the learning abilities of castrated rats, and whether food or water was more likely to inspire a rat to escape. Stone was clinical, systematic, and cautious to his bone marrow. He was widely respected as a meticulous observer who built his scientific cases detail by solid detail.

He and Harry were a near perfect mismatch of temperaments.

Stone used to tell his students that good researchers "will push the domain of science forward inch by inch." Harry hated the thought. He wanted to leap. Never mind inch by inch, Harry used to pun; his professor was going to pursue scientific inquiry stone by stone. Stone expected only orderly science. Another of his Ph.D. students, William Mason, who would later do postgraduate work with Harry, recalls doing a study for Stone and, being in a hurry, hastily scribbling his findings on whatever piece of paper he could find. Stone, frowning, called him aside: "Mason, we do not record data on scraps."

Years later, Harry hadn't forgotten an encounter with Stone when "I was almost bleeding to death from a lab accident and met him in the hall." Stone promptly began a detailed discussion of an experiment, describing apparatus design and testing plans while Harry "wondered how long it would be before he would notice the blood all over my hand and my gown. Finally, he looked down and said, 'Oh, bitten by a rat, eh?' You see, he was methodical; he wasn't jarred by the fact that a person was bleeding to death."

There's no doubt, anyway, that Stone would never have spun a small rat bite into a near-death injury. Things were what they were.

And if he didn't teach the habit of storytelling out of Harry, he did teach him a lifelong respect for doing the research properly, for lining up facts with precision. Stone's students agree that even if he was chilly personally, he radiated a love of good science. He did his best to teach that, too. Harry and his professor maintained respectful relations; when Stone retired as editor of the *Journal of Comparative and Physiological Psychology,* he successfully recommended that Harry take over the job. When Stone died, it was Mason who wrote the professional tribute and Harry who encouraged him to "use some of that Lincolnesque style of yours, Mason," in praising their former professor. Years later, Harry was still joking about Stone and the rat bite incident, telling a magazine interviewer that his old professor was basically a good man and that "he probably went out and bawled the hell out of the rat."

Stone directed Harry's Ph.D. dissertation, a 170-page exploration of feeding habits in baby rats. The study was classic Stone, completely and obsessively thorough about what infant rodents liked to drink, when, where, and how. Harry was polite in thanking Stone for his "consideration, his suggestions and his consistently stimulating interest in this investigation." But one of Harry's fellow students, psychologist Robert Sears, suggested that the dissertation fostered a dislike of rat research that Harry never overcame. It was those hours on a "pedestrian rat problem," under Stone's guidance, that "soured [Harry] forever on both rats" and statistical analysis, according to Sears.

Harry concurred. He used to say that he'd seen enough of rats at Stanford—in Stone's lab, in Miles's garage—to last him a lifetime. "Although I am thought of as a monkey psychologist, I'm sure that I have spent more man-hours studying rats than any two living psychologists combined." He announced that when he took over as journal editor, he was more than ready to resist if "somebody tried to push a rat paper down my throat." For the rest of his life, he insisted on calling psychological studies with rats "rodentology."

Still, Harry Harlow's future glimmers in that dissertation, once you get beyond the title: "An Experimental Study of the Feeding

Reactions and Related Behavior Patterns of the Albino Rat." The primary discovery is, as Sears pointed out, no real surprise. All those hours of research showed that rats will swallow liquids other than rat milk as long they think the taste half-way decent. If it tastes bad, they'd just as soon spit it out. The rats in Harry's study would accept whole and diluted cow's milk and sugar solutions. If nothing else was available, they would reluctantly make do with orange juice and even cod liver oil. The bitter taste of quinine, the sting of a weak acid solution, and the sharpness of salt solutions produced instant rejection—which meant spitting it out and squirming to get away.

Perhaps more to the point, Harry began to learn that the baby rats needed constant "mothering," including guidance in how much food they should take. In his first cows' milk test, he fed the rats every three hours, which turned out to be not nearly enough. One of his little rats died of malnutrition. In dismay, he doubled the feeding schedule. This turned out to be too much. The baby rats happily sucked down all the milk but, by the tenth day, all of them were dead from overfeeding. Being a parent—even the scientific surrogate for a lactating rat mother—clearly required knowledge and experience, including when to say "Enough." It also raised another question.

Are there conditions that inhibit feeding, that simply turn off all that natural greed and hunger? Harry tried some simple experiments in temperature. Rat families were placed on a glass floor, which could be alternatively chilled with ice cubes or warmed by an electric heating pad. He discovered that too much cold simply froze the feeding process. If they were chilly, the little rats just wouldn't eat. It was as if they were numbed to a standstill. Curiously, though, warming the floor didn't improve their feeding habits, either. The baby rats were likely to just huddle down into the warmth. They needed to be cared for, coaxed by something more than the ambient temperature. Mother rats, as it turns out, squash their infants firmly between their own bodies and the nest while the babies eat. The warmth, the sense of being wedged into a big family pancake of sorts, seems to help stir up the hunger response. Scientists could manipulate eyedroppers and drip milk and juice and sugar-water down the throats of baby rats, but glass instruments weren't nearly as productive as the simple act of being sat on by a mother rat.

The next set of experiments was not pedestrian at all, although it's not clear that anyone involved really appreciated the potential. Neither Harry nor Stone followed up on the results. They were, though, a haunting testament to mother nature. Harry built a device in which mothers and baby rats were separated by a mesh barrier with small holes cut into it, large enough for the newborn rats to squeeze through, but not the mothers. Lost and bewildered, on the wrong side of the mesh, the babies crawled in aimless circles. The mother rats, on the other hand, weren't aimless at all. They were desperate to get to their pups. They would bite the mesh angrily, try to force their way through the too-small holes; and when the barrier was removed, they immediately began collecting the young. Even if the mothers were hungry, even if food was placed temptingly before them, they would first gather their families to safety. Then they would eat.

What lay behind the intensity of this response, the imperative riptide pull of mother toward child? Was it a simple sensory reflex? At Stone's direction, Harry removed ovaries, blinded the female rats, and removed their olfactory bulbs. Sightless, hormone-deprived—it didn't matter. The mother rats crawled determinedly toward the baby rats. They were slower, maybe, but the homing instinct was magnetic, needle to the north.

On the title page of Harry's dissertation, directly under that stuffy title, is one more, very different clue about the author's future direction. The paper is credited not to Harry Frederick Israel of Fairfield, Iowa, but to Harry Frederick Harlow of Palo Alto, California. And, to understand that change—the disappearance of *Israel über alle*—one needs to appreciate both the strength of Harry's dreams and the extraordinary presence and influence of Lewis M. Terman.

Terman was a luminary in the still new field of psychology. He knew it, his colleagues knew it, the university knew it. Let him fall ill and the Stanford administration paid anxious attention. In 1926, when Terman canceled a trip to the East Coast due to influenza, the university president, Ray Lyman Wilbur, responded with a solicitous note: "I am sorry to learn that you have not been entirely well, but am glad that you are taking care of yourself." At Stanford in the 1920s, Terman wasn't just a famous and innovative researcher, he was also a powerful one. It was *his* psychology department and everyone—down to the lowliest student—knew that.

To paint Terman as pure autocrat would be misleading. Like Miles, he considered his students an extended family and he paid attention to them. He could be disarmingly affectionate. He and his wife, Anna, visited a graduate student, Jessie Linton, in the hospital after she had given birth to her first child. They both demanded to hold the baby. Linton recalled teasing her professor, saying she thought men didn't like to be handed small, squirmy infants. "That's what you think," Terman replied, cuddling the child to him. He would take students on picnics to celebrate their achievements. He held weekly seminars at his house, open to undergraduate students if they were interested. He charmed and he listened and he prodded and if he saw any promise in you at all, he would push you relentlessly to exceed. "Terman was entirely different from Stone," Harry said, "He was out to find the creative and he took great pride in that."

By the time Harry Israel arrived at Stanford, Terman was in his mid–forties, his red hair flecked with gray, his face wonderfully rumpled, his health uncertain, his vision straight ahead. Terman's particular research focused on human intelligence. Tests to "measure" intelligence had begun to appear in the late nineteenth century, both in the United States and Europe; many psychologists believed that such examinations were yet another way to demonstrate that their field was growing into a precise, documented, quantifiable branch of science.

Terman used the intelligence test as a probe, a research tool to assess human potential. He had adapted the test for the purpose. An earlier version, created by French psychologist Alfred Binet, had been more of a teacher's aid. Binet saw his test as a way to pick out children who needed extra tutoring, to better tailor their schooling to their needs. But Terman saw it differently; less compassionately, maybe, and more clinically. Terman refocused the exam into a purer test of analytical talent. The improved version measured such things as one's ability to think through the angles of a triangle or solve that well-known problem of two trains approaching a station at different speeds. Terman had little interest in judging whether students were being taught properly. He cared about their native intelligence, their innate capability to reason through a challenging problem. He did hope that his test would someday allow society to sort people by their abilities. Perhaps children could then be taught in accordance with their talents. That way, the brightest could be made even brighter. But he didn't believe that improving teaching was the primary issue because, frankly, he believed people were born smart—or were not.

His adaptation of Binet's test would become known as the Stanford-Binet. It is still the granddaddy of all IQ and scholastic aptitude tests used today. Under Terman's design, the Stanford-Binet sorted a person into one of four categories: gifted, bright, average, or special. There was a range of ability in each of those groups. On the Stanford-Binet scale, if one scored below 30, that indicated a drooling, shuffling kind of mental handicap. A person had to rise into the 70s before the numbers shifted more toward intelligence. A score between 70 and 79 was still considered borderline retardation—what psychologists of the time called the "feebleminded." In other words, 79 and down put you in the "special" group. Basic competence—being average—emerged in the 80s. At about 100, one started creeping into the "bright" region, and brilliance, or "being gifted," began at a score of 140 or so.

Today, IQ testing is regarded by many as a limited probe, a measure primarily of analytical abilities. In retrospect, many psychologists also acknowledge that Terman and his colleagues in the IQ arena could sound elitist—and worse. The word "moron" was coined by another believer in intelligence testing, Henry Goddard, who used it to describe low scorers. Goddard went on to speak virulently against immigration, insisting that Jewish and Eastern European immigrants would dilute good Northern European stock with their "low-intellect" genes. Supporters of intelligence testing argued, successfully, that "feebleminded" men and women should be sterilized to avoid reproducing additional generations of imbeciles. Terman himself wrote that genetic superiority could be expected to predict social superiority.

But Terman was also willing to ask hard questions of the so-called elite. For instance, did very smart people naturally rise to the top, the cream floating up over the milky rest of the population? Or did they need extra support to rise? A few years before Harry Israel came to Stanford, Terman began a long-term study of the gifted. He started with exceptional students who were found first by questionnaires sent to elementary school teachers. Then he ran those students and their siblings through IQ tests. All the children that Terman selected scored at least 140 on the Stanford-Binet scale and some as high as 192. His core group—363 boys, 313 girls—had to pass other tests as well.

Because Terman thought gifted children should perform well in real life as well as on paper, he screened against handicaps such as shyness and disabilities such as limping or stuttering. His questionnaire asked about "prudence, forethought, willpower, humor, cheerfulness, fondness of large groups, popularity, generosity, truthfulness, commonsense, and energy." He looked for children who had a desire to excel. And just in case those filling out the form were unsure what such a desire was, Terman provided a definition: "Does his utmost to stand first."

There was no doubt that self-confidence was the order of the day when Harry was at Stanford. Terman expected his chosen students to damn well be smart. And act it. He selected carefully. One favorite was Nancy Bayley, who would become one of UC Berkeley's best-known child psychologists and whose own work on cognitive development would eventually directly contradict Terman's. (Bayley showed that parenting styles *did* seem to affect IQ. Little boys raised by unaffectionate mothers showed steady erosion in test scores. Little girls also faltered, especially if they were harshly restricted and disciplined.) Terman did not expect his students to agree with him on everything. He did expect them to be good scientists, and really good if they wanted to win their arguments. Bayley credited Terman for teaching her to be a perfectionist. He was, she said, meticulous in his own work, always ready to praise students when they did well, and "very critical of sloppy work."

Another graduate student recalled spending a year working on his dissertation, only to be told by Terman that it was substandard and he would have to begin again. Terman insisted that real scientists never took time off. Even at this exalted stage of his career, he often worked late into the night: "He was always working near the limits of knowledge," said psychologist Robert Bernreuter, another of Terman's protegés. And Terman wanted his students to venture over those limits, too. "Usually Terman would point out two or three times each seminar something that needed additional research. This caused us to develop both a profound respect for research, and the feeling that we should do something about it," Bernreuter said.

Terman's students wanted desperately to "do something about it" in a way that would gain his approval. Perhaps more than some of the other students, Harry doubted his ability to impress the master. He knew he was smart, creative—even funny on occasion—but could he demonstrate that while he was at Stanford? The school brought out all the tentativeness and shyness you might expect from the son of a failed doctor in rural Iowa. His description of himself at that time was of "a shy, retiring youth with a rather poetic outlook on life. I tended to apologize to doors before opening them." And when he did apologize, there was a slight speech defect that would have undoubtedly stricken him from Terman's study of exceptional students. Pronouncing the letter "r" had caused Harry trouble since his childhood and sometimes gave his conversation a cartoonish quality, in the "silly wabbit" style of Elmer Fudd. Embarrassed, he often chose to say nothing rather than to sound goofy. On the grounds of shyness, on the grounds of his speech defect, he could not have entered Terman's gifted study. And Terman made that almost painfully clear to him.

When Harry started on his master's degree, "Dr. Terman called me into his office and told me he thought I was a bright young man but that I was so timid that I would never be able to speak in public." The "r's" only made that problem worse. Terman "recommended strongly that my future lay in teaching in a junior college as I would never be able to speak effectively before really large audiences." Terman even had his secretary check into the requirements for such a job. It turned out that teaching at a junior college required education courses that Harry didn't have. "As a result, I was condemned to get a Ph.D."

Upon Harry's graduation in 1930, Terman called him back. He was still worrying about Harry's future. This time it had do with the negative consequences of his last name. "He said that since my name was Israel, they had found it impossible to place me in an adequate academic position because of anti-Jewish prejudice." Walter Miles had been talking Harry up at other universities and had been asked constantly about his student's religious background. The dean of a large state university told Miles that he didn't care how good the young psychologist was, he was not going to hire anyone with the last name of Israel. Harry often looked back with real disbelief at the depth of discrimination in the 1930s, even on supposedly enlightened university campuses. "I don't want to imply that I was persecuted, because I wasn't; but with the name Israel, and because I was a timid boy, I certainly had seen discrimination." The Israels were not Jewish: "Gentile for generations. An aunt traced the name back to 1753, and found an ancestor who had been buried in a Jewish cemetery." Harry had no patience with anti-Semitism; when he told the story of the name change, he tried to make that perfectly clear, even proposing that the faint Jewish ancestry was responsible for any intelligence in the Israel line: "I often wondered where the family got any brains."

Terman said it didn't matter whether Harry was Jewish or not; the problem was that his name sounded Jewish. "He also indicated that even though the Depression had already hit they would keep me on some kind of basis for the forthcoming year." As it turned out, Harry didn't need the extra support from Stanford. Shortly after his conversation with Terman—while the young psychologist was still considering his department head's proposal—a job offer came through. The University of Wisconsin had sent a one-line telegram to Harry Israel. It asked, "Will you accept an assistant professorship at the UW paying $2,750 a year?" In a heartbeat. He was packed, on the road, out of there. He almost left California, yet, as Harry Israel.

But Harry's last name still troubled Terman. In his letter to Wisconsin recommending Harry Israel, Terman had acknowledged the Jewish sound of the name. He then assured the potential employers that Dr. Israel was not "*that* kind of Jew." He called Harry into the office and said that he was glad about the job but he still thought the name Israel was just wrong, too negative. It would continue to hold him back. Didn't Harry want to have a great future, not just an ordinary one? Of course he did. This was the Harry Israel who had written down "fame" as his ambition in his high school yearbook. He still desperately wanted to please Terman in some way, to prove himself beyond that junior college designation. And "I had seen anti-Jewish prejudice and did not want any son or particularly daughter of mine to go through it." Okay, Harry said, give me a new name. Terman suggested that Harry choose a name that at least belonged to his family. Harry came up with two possibilities: Crowell, after an uncle; and Harlow, from his father's middle name. "Terman chose Harlow and, as far as I know, I am the only scientist who has ever been named by his major professor."

By the time the news reached Fairfield, Stanford was already printing the graduation program, and listed under Ph.D. graduates was that new man again, Harry F. Harlow of Palo Alto, California. Only once, and that was to a close friend, Harry said that he regretted the change, that it seemed to dishonor whatever Jewish ancestry he had, be it only 1⁄64th that ran in the Israel family. That was long after he'd left Stanford, and after World War II and Adolf Hitler had put an end to the fancy—if people ever really believed it—that there could be benign prejudice toward any people, that such attitudes were mere silliness. His fall-back position—as always—was a joke. In an interview in *Psychology Today,* he told it like this: "So I became a Harlow. I guess I'm not alone. Once a man called me up and said he was looking up the Harlow ancestry. I said I was sorry, but I had changed my name." "Oh, heavens, not again," he replied. "Everyone named Harlow that is worth a damn has changed his name."

Changing the name, of course, doesn't change the person. Years later, one of Harry's best-known post-doctoral researchers, California psychologist William Mason, would wonder which man was the real Harry: the Wisconsin crusader called Harry Harlow or the shy loner from Iowa named Harry Israel. "What was the real man like?" Mason asked. "Very complex." There's one aspect of the almost forgotten Harry Israel, though, that remains straightforward: He understood that you could win a lost cause. Against both odds and expectations, he'd become a Stanford-trained research psychologist. During his career, that belief—that you should rarely declare a battle lost—would guide many of his most defiant research choices. Eventually, his fondness for unpopular causes would lead him to labor for love. And to appreciate what a lost cause that was in the world of mid-twentieth-century psychology, you must appreciate the depth and righteousness of the opposition to love as part of daily life. Psychologists argued vehemently against cuddling children. Doctors stood against too close contact between even parent and child. There was real history behind this, built on experience from orphanages and hospitals, built on lessons learned from dead children and lost babies. There were careful experiments and precise data and numerical calculations of behavior to prove that emotions were unnecessary and unimportant. There was a decades-thick wall of research and an army of researchers to counter any upstart psychologist, including Harry Harlow. Naturally, it was just the kind of challenge that appealed to him most.

TWO

Untouched by Human Hands

The apparent repression of love by modern psychologists stands in sharp contrast with the attitude taken by many famous and normal people.

Harry F. Harlow,
***The Nature of Love,* 1958**

THE FRUSTRATING, IMPOSSIBLE, TERRIBLE THING about orphanages could be summarized like this: They were baby killers.

They always had been. One could read it in the eighteenth-century records from Europe. One foundling home in Florence, The Hospital of the Innocents, took in more than fifteen thousand babies between 1755 and 1773; two thirds of them died before they reached their first birthday. In Sicily, around the same time, there were so many orphanage deaths that residents in nearby Brescia proposed that a motto be carved into the foundling home's gate: "Here children are killed at public expense." One could read it in the nineteenth-century records from American orphanages, such as this report from St. Mary's Asylum for Widows, Foundlings, and Infants in Buffalo, New York: From 1862 to 1875, the asylum offered a home to 2,114 children. Slightly more than half—1,080—had died within a year of arrival. Most of those who survived had mothers who stayed with them. "A large proportion of the infants, attempted to be raised by hand, have died although receiving every possible care and attention that the means of the Sisters would allow as to food, ventilation, cleanliness, etc."

And yet babies, toddlers, elementary school children, and even adolescents kept coming to foundling homes, like a ragged, endless, stubbornly hopeful parade. In the orphanages, the death of one child always made room for the next.

Physicians were working in and against an invisible lapping wave of microorganisms, which they didn't know about and couldn't understand. Cholera flooded through the foundling homes, and so did diphtheria and typhoid and scarlet fever. Horrible, wasting diarrheas were chronic. The homes often reeked of human waste. Attempts to clean them foundered on inadequate plumbing, lack of hot water, lack even of soap. It wasn't just foundling homes, of course, where infections thrived in the days before antibiotics and vaccines, before chlorinated water and pasteurized milk. In the United States, more than one fourth of the children born between 1850 and 1900 died before age five. But foundling homes concentrated the infections and contagions, brought them together in the way a magnifying glass might focus the sun's rays until they burn paper. The orphanages raised germs, seemingly, far more effectively than they raised children. If you brought a group of pediatricians together, they could almost immediately begin telling orphanage horror stories—and they did.

In 1915, a New York physician, Henry Chapin, made a report to the American Pediatric Society that he called "A Plea for Accurate Statistics in Infants' Institutions." Chapin had surveyed ten foundling homes across the country; his tally was—by yesterday's or today's standards—unbelievable. At all but one of the homes, every child admitted was dead by the age of two. His fellow physicians rose up—not in outrage but to go him one better. A Philadelphia physician remarked bitterly that "I had the honor to be connected with an institution in this city in which the mortality among all the infants under one year of age, when admitted to the institution and retained there for any length of time, was 100 percent." A doctor from Albany, New York, disclosed that one hospital he had worked at had simply written "condition hopeless" on the chart as soon as a baby came into the ward. Another described tracking two hundred children admitted into institutions in Baltimore. Almost 90 percent were dead within a year. It was the escapees who mostly survived, children farmed out to relatives or put in foster care. Chapin spent much of the rest of his career lobbying for a foster care system for abandoned children. It wasn't that he thought foster homes would necessarily be kinder or warmer—he hoped only that they wouldn't kill children so quickly.

By Chapin's time, of course, thanks to researchers such as Louis Pasteur and Alexander Fleming and Edward Jenner, doctors recognized that they were fighting microscopic pathogens. They still didn't fully understand how those invisible infections spread—only that they continued to do so. The physicians' logical response was to make it harder for germs to move from one person to the next. It was the quarantine principle: Move people away from each other, separate the sick from the healthy. That principle was endorsed—no, loudly promoted—by such experts of the day as Dr. Luther Emmett Holt, of Columbia University. Holt made controlling childhood infections a personal cause. The premier childcare doctor of his time, he urged parents to keep their homes free of contagious diseases. Remember that cleanliness was literally next to Godliness. And remember, too, that parents, who weren't all that clean by doctors' standards, were potential disease carriers. Holt insisted that mothers and fathers should avoid staying too close to their children.

Before Holt, American parents usually allowed small children to sleep in their bedrooms or even in their beds. Holt led a crusade to keep children in separate rooms; no babies in the parental bedroom, please; good childcare meant good hygiene, clean hands, a light touch, air and sun and space, including space from you, mom and dad. And that meant avoiding even affectionate physical contact.

What could be worse than kissing your child? Did parents really wish, asked Holt, to touch their baby with *lips,* a known source for transmitting infection?

If parents had doubts about such lack of contact, Holt's colleagues did not. In the 1888 *The Wife's Handbook* (with Hints on Management of the Baby), physician Arthur Albutt also warned each mother that her touch could crawl with infection. If she really loved the baby, Albutt said, she should maintain a cautious distance: "It is born to live and not to die" and so always wash your hands before touching, and don't "indulge" the baby with too much contact so that "it"—the baby is always "it" in this book—may grow up to fill a "useful place in society."

In foundling homes, wedged to the windows with abandoned children, there was no real way to isolate an ailing child—nor did anyone expect the foundlings to occupy many useful places in society. But administrators did their best to keep their charges alive. They edged the beds farther apart; they insisted that, as much as possible, the children be left alone. On doctors' orders, the windows were kept open, sleeping spaces separated, and the children touched as little as possible—only for such essentials as a quick delivery of food or a necessary change of clothes. A baby might be put into a sterile crib with mosquito netting over the top, a clean bottle propped by its side. The child could be kept virtually untouched by another human being.

In the early twentieth century, the hyperclean, sterile-wrapped infant was medicine's ideal of disease prevention, the next best thing to sending the baby back to the safety of the womb. In Germany, physician Martin Cooney had just created a glass-walled incubator for premature infants. His *Kinderbrutanstalt* ("child hatchery") intrigued both manufacturers and doctors. Because preemies always died in those days anyway, many parents handed them over to their physicians. Doctors began giving them to Cooney. He went on an international tour to promote the hatchery, exhibiting his collection of infants in their glass boxes. Cooney went first to England and then to the United States. He showed off his babies in 1902 at the Pan American Exposition in Buffalo, New York. During the next two years, he and his baby collection traveled to shows as far west as Nebraska. Cooney settled in Coney Island, where he successfully cared for more than five thousand premature infants. Through the 1930s, he continued, occasionally, to display them. In 1932, he borrowed babies from Michael Reese Hospital for the Chicago World's Fair and sold tickets to view the human hatchlings. According to fair records, his exhibit made more money that year than any other, with the exception of that of Sally Rand, the famed fan-dancer. The babies in the boxes were like miracles of medicine; they were alive when generations before them had died. Cooney said his only real problem was that it was so hard to convince mothers to take their infants back. Oddly enough, they seemed to feel disconnected from those babies behind the glass.

Sterility and isolation became the gods of hospital practice. The choleras and wasting diarrheas and inexplicable fevers began to fall away. Children still got sick—just not so mysteriously. There were always viruses (measles, mumps, things we now vaccinate against) and still those stubborn bacterial illnesses that plague us today: pneumonias, respiratory infections, drearily painful ear infections. But, now, doctors took the position that even the known infections could be best handled by isolation. Human contact was the ultimate enemy of health. Eerily unseeable pathogens hovered about each person like some ominous aura. Reports from doctors at the time read like descriptions of battle zones in which no human was safe—and everybody was dangerous. One such complaint, by Chicago physician William Brenneman, discussed the risks of letting medical personnel loose in the wards. Nurses weren't allowed enough sick leaves and they were bringing their own illnesses into the hospital; interns seemed to not appreciate that their "cold or cough or sore throat" was a threat. Physicians themselves, Brenneman added sarcastically, apparently felt they were completely noninfectious when ill, as long as they wore a long "white coat with black buttons all the way down the front." How could you keep illness out of hospital when doctors and nurses kept coming in?

Brenneman, of Children's Memorial Hospital in Chicago, thought children's wards were similar to concentration camps, at least when it came to infection potential. He evoked the prison camps of World War I, where doctors had found that captured soldiers were crawling with streptococcus bacteria. Were wards so different? Tests had shown that 105 of 122 health workers at the hospital were positive for the same bacteria, a known cause of lethal pneumonias. "It is known what the streptococcus did in concentration camps during the World War. One is constantly aware of what it does in the infant ward under similar conditions of herding and massed contact." The less time a child spent in the hospital the better was Brenneman's rule, and he urged doctors to send their patients home; or if they had no home, into foster care, as quickly as possible. And if they had to be hospitalized? Push back the beds; wrap up the child quickly, keep even the nurses away when you could.

Harry Bakwin, a pediatrician at Bellevue in New York, described the children's ward of the 1930s like this: "To lessen the danger of cross infections, the large open ward of the past has been replaced by small, cubicled rooms in which masked, hooded, and scrubbed nurses and physicians move about cautiously so as not to stir up bacteria. Visiting parents are strictly excluded, and the infants receive a minimum of handling by the staff." One hospital even "devised a box equipped with inlet and outlet valves and sleeve arrangements for the attendants. The infant is placed in this box and can be taken care of almost untouched by human hands." By such standards, the perfectly healthy child would be the little girl alone in a bed burnished to germ-free perfection, visited only by gloved and masked adults who briskly delivered medicine and meals of pasteurized milk and well-washed food.

Hospitals and foundling homes functioned, as Stanford University psychologist Robert Sapolsky puts it today, "at the intersection of two ideas popular at the time—a worship of sterile, aseptic conditions at all costs, and a belief among the (overwhelmingly male) pediatric establishment that touching, holding and nurturing infants was sentimental maternal foolishness." It wasn't just that doctors were engaged in a quest for germ-free perfection. Physicians, worshipping at the altars of sterility, found themselves shoulder-to-shoulder with their brethren who studied human behavior. Their colleagues in psychology directly reassured them that cuddling and comfort were bad for children anyway. They might be doing those children a favor by sealing them away behind those protective curtains.

Perhaps no one was more reassuring on the latter point than John B. Watson, a South Carolina–born psychologist and a president of the American Psychological Association (APA). Watson is often remembered today as the scientist who led a professional crusade against the evils of affection. "When you are tempted to pet your child remember that mother love is a dangerous instrument," Watson warned. Too much hugging and coddling could make infancy unhappy, adolescence a nightmare—even warp the child so much that he might grow up unfit for marriage. And, Watson warned, this could happen in a shockingly short time: "Once a child's character has been spoiled by bad handling, which can be done in a few days, who can say that the damage is ever repaired?"

Nothing could be worse for a child, by this calculation, than being mothered. And being mothered meant being cradled, cuddled, cosseted. It was a recipe for softness, a strategy for undermining strong character. Doting parents, especially the female half of the partnership, endowed their children with "weaknesses, reserves, fears, cautions and inferiorities." Watson wrote an entire chapter on "The Dangers of Too Much Mother Love," in which he warned that obvious affection always produced "invalidism" in a child. The cuddling parent, he said, is destined to end up with a whiny, irresponsible, dependent failure of a human being. Watson, who spent most of his research career at Johns Hopkins University, was a nationally known and respected psychologist when he trained his sights on mother love. Articulate, passionate, determined, he was such an influential leader in his field, that his followers were known as "Watsonian psychologists." And like him, they came to consider coddling a child as the eighth of humankind's deadly sins. "The Watsonian psychologists regard mother love as so powerful (and so baneful) an influence on mankind that they would direct their first efforts toward mitigating her powers," wrote New York psychiatrist David Levy in the late 1930s.

Watson believed that emotions should be controlled. They were messy; they were complicated. The job of a scientist, of any rational human being, should be to figure out how to command them. So he was willing to study emotions, but mostly to show that they were as amenable to manipulation as any other basic behavior. The emotion of rage, he said, could be induced in babies by pinning them down. That was a simple fact, observable and measurable and controlled by the mastery of science. If it sounds cold, he meant it to be. Watson, as many of his colleagues, was driven by a need to prove psychology a legitimate science—with the credibility and chilly precision of a discipline such as physics.

Psychology was a young science at the time, founded only in the nineteenth century. Until that point—perhaps until Darwin—human behavior was considered the province of philosophy and religion. Scientists considered physics, astronomy, chemistry as serious research subjects, but those disciplines had hundreds of years behind them. Even one of the founders of the American Psychological Association, William James of Harvard, said that psychology wasn't a science at all—merely the hope of one.

As a child, Watson had been dragged to tent revival after tent revival by his mother. He still remembered with revulsion the sweaty intensity of the faithful. He was determined to wash the remnants of spirituality and, yes, emotion out of his profession. "No one ever treated the emotions more coldly," Harry Harlow would say years later. To his contemporaries, Watson only argued that a scientific psychology was the way to build "a foundation for saner living." He proposed stringent guidelines for viewing behavior in a 1913 talk still known as the Behaviorist Manifesto.

"Psychology as the behaviorist views it is a purely objective, experimental branch of natural science," he insisted. Its goal was the prediction and control of behavior. "Introspection forms no essential part of its methods, and neither does consciousness have much value." Psychologists should focus on what could be measured and modified. In the same way that animals could be conditioned to respond, so could people. The principle applied most directly to children. Watson's psychology was in near perfect opposition to the intimate, relationship-focused approach that Harry Harlow would develop. Rather, he argued that adults—parents, teachers, doctors—should concentrate on conditioning and training children. Their job was to provide the right stimulus and induce the correct response.

And that was what Watson demanded, forcefully, in his 1928 bestseller, *The Psychological Care of the Child and Infant.* The British philosopher Bertrand Russell proclaimed it the first child-rearing book of scientific merit. Watson, he said, had triumphed by studying babies the way "the man of science studies the amoebae." *The Atlantic Monthly* called it indispensable; the *New York Times* said that Watson's writings had begun "a new epoch in the intellectual history of man." *Parents* magazine called his advice a must for the bookshelf of every enlightened parent.

From today's perspective, it's clear that Watson had little patience for parents at all, enlightened or not. Watson wrote that he dreamed of a baby farm where hundreds of infants could be taken away from their parents and raised according to scientific principles. Ideally, he said, a mother would not even know which child was hers and therefore could not ruin it. Emotional responses to children should be controlled, Watson insisted, by using an enlightened scientific approach. Parents should participate in shaping their children by simple, objective conditioning techniques. And if parents chose affection and nurturing instead, ignoring his advice? In his own words, there are "serious rocks ahead for the over-kissed child." Watson demanded not only disciplined children but disciplined parents. His instructions were clear: Don't pick them up when they cry; don't hold them for pleasure. Pat them on the head when they do well; shake their hands; okay, kiss them on the foreheads, but only on big occasions. Children, he said, should be pushed into independence from the day of their birth. After a while, "you'll be utterly ashamed of the mawkish, sentimental way you've been handling your child."

Watson was a hero in his own field, hailed for his efforts to turn the soft-headed field of psychology into a hard science. He became a hero in medicine because his work fit so well with the "don't touch" policies of disease control. The physicians of the time also considered that affection was, well, a girl thing, something to be sternly controlled by men who knew better. The *Wife's Handbook* flatly warns mothers that their sentimental natures are a defect. The book's author, Dr. Arthur Albutt, takes a firm stand against spoiling, which he defines as picking babies up when they cry, or letting them fall asleep in one's arms. "If it cries, never mind it; it will soon learn to sleep without having to depend on rocking and nursing." Dr. Luther Holt took the same stance and his publication, *The Care and Feeding of Children,* was an even bigger success. There were fifteen editions of his book between 1894 and 1935. Holt believed in a rigorous scientific approach to the raising, or let's say, taming of the child. The whole point of childhood was preparing for adulthood, Holt said. To foster maturity in a child, Holt stood against the "vicious practice" of rocking a child in a cradle, picking him up when he cried, or handling him too often. He urged parents not to relax as their child matured. Holt was also opposed to hugging and overindulging an older child.

It's easy today to wonder why anyone would have listened to this paramilitary approach to childcare. Undoubtedly—or at least we might hope—plenty of parents didn't take heed. Yet, Holt and Watson and their contemporaries were extraordinarily influential. Their messages were buoyed by a new, almost religious faith in the power of science to improve the world. The power of technology to revolutionize people's lives was a tangible, visible force. Gaslights were flickering out as homes were wired for electricity. The automobile was beginning to sputter its way down the road. The telegraph and telephone were wiring the world. There were mechanical sewing machines, washing machines, weaving machines—all apparently better and faster than their human counterparts. It was logical to assume that science could improve we humans as well.

John Watson wasn't the only researcher to publicly urge scientific standards for parenting. The pioneering psychologist G. Stanley Hall, of Clark University, entered the childcare field as well. In 1893, Hall helped found the National Association for the Study of Childhood. His own work focused on adolescence and he believed that the difficulties encountered at this time of life were in part due to mistakes by parents and educators in the early years. Hall admired much about what he called the adolescent spirit and its wonderfully creative imagination. But it needed discipline, he said, moral upbringing, strict authority to guide it.

Speaking to the National Congress of Mothers—a two-thousand-member group organized in 1896 to embrace the concept of scientific motherhood—Hall urged Victorian tough love upon them. Their children needed less cuddling, more punishment, he said; they needed constant discipline. After Hall's talk to the mothers' congress, the *New York Times* rhapsodized in an editorial, "Given one generation of children properly born and raised, what a vast proportion of human ills would disappear from the face of the Earth." Women at the conference left determined to spread the word. No more adlibbing of childcare, they insisted. There were real experts out there, men made wise by science. Parents needed to pay attention. "The innocent and helpless are daily, hourly, victimized through the ignorance of untrained parents," said the Congress of Mothers' president, Alice Birney, in 1899. "The era of the amateur mother is over." (The mothers' congress, by the way, changed and grew and eventually became part of the PTA.)

The demand for scientific guidance was so pressing that the federal government's Child Bureau—housed in the Department of

Labor—after all, childrearing was a profession—went into the advice business. The bureau recruited Luther Holt as primary advisor on its "Infant Care" publications. Between 1914 and 1925, the Labor Department distributed about 3 million copies of the pamphlet. Historian Molly Ladd-Taylor, in her wonderfully titled book, *Raising a Baby the Government Way,* reports that the Child Bureau received up to 125,000 letters a year asking for parenting help. The bureau chief, Julia Lathrop, said that each pamphlet was "addressed to the average mother of this country." The government was not, she emphasized, trying to preempt doctors. "There is no purpose to invade the field of the medical or nursing professions, but rather to furnish such statements regarding hygiene and normal living as every mother has a right to possess in the interest of herself and her children."

The "Infant Care" pamphlet covered everything from how to make a swaddling blanket to how to register a birth. It discussed diapers, creeping pens (which we today call playpens), meals from coddled eggs to scraped beef, teething, nursing, exercise, and, oh yes, "Habits, training, and discipline." After all, "the wise mother strives to start the baby right."

The care of a baby—according to the federal experts—demanded rigid discipline of both parent and child. Never kiss a baby, especially on the mouth. Do you want to spread germs and look immoral? (This part, obviously, straight from the mouth of Luther Holt.) And the government, too, wanted to caution mothers against rocking and playing with their children. "The rule that parents should not play with the baby may seem hard, but it is without doubt a safe one." Play—tickling, tossing, laughing—might make the baby restless and a restless baby is a bad thing. "This is not to say that the baby should be left alone too completely. All babies need 'mothering' and should have plenty of it." According to federal experts, mothering meant holding the baby quietly, in tranquility-inducing positions. The mother should stop immediately if her arms feel tired. The baby is never to inconvenience the adult. An older child—say above six months—should be taught to sit silently in the crib; otherwise, he might need to be constantly watched and entertained by the mother, a serious waste of time in the opinion of the authors. Babies should be trained from infancy, concludes the pamphlet, so "smile at the good, walk away from the bad—babies don't like being ignored."

Universities also began offering scientific advice to untutored parents. Being research institutions, they tended to reflect John Watson and the zeitgeist of experimental psychology. Reading them today is curiously like reading a pet-training guide—any minute, the mother will be told to issue a "stand-stay" command to her toddler. In the *Child Care and Training* manuals, published by the University of Minnesota's Institute of Child Welfare, the authors advised that the word "training" refers to "conditioned responses." They assured their readers that when a mother smiles at a baby, she is simply issuing a "stimulus." When the baby smiles back, he is not expressing affection. The baby has only been conditioned to "respond" to the smile.

Further, parents should be aware that conditioning is a powerful tool, the Minnesota guidebook warned. For instance, if a child falls down and hurts herself, mothers and fathers should not condition her to whine. They might do that if they routinely pick her up and comfort her. Treat injury lightly and "tumbles will presently bring about the conditioned response of brave and laughing behavior," the guidebook advised. Watson had declared that babies feel only three emotions: fear, rage, and love (or the rudiments of affection), and the Minnesota psychologists agreed. They warned that it is easy to accidentally condition unwanted fears. The researchers cited the common practice of locking children in a dark room to punish them. They recommended against it. This, they said, only conditions the child to fear darkness. A stern word, a swift swat, is so much better. The scientists also suggested that parents try not to worry about their children and their safety so much: Fear conditions fear. "The mother who is truly interested in bringing up children free of fear will try to eliminate fear from her own life." Watson equated baby love with pleasure, brought on by stroking and touch. But he also believed that too much of such affection would soften the moral fiber of the children. So did the Minnesota group. Their manual states that although ignoring and being indifferent to a child *could* cause problems, it was "a less insidious form of trouble than the over-dependence brought about by too great a display of affection."

It was serendipity, it was timing—the ideas fit together like perfectly formed pieces of a puzzle. Medicine reinforced psychology; psychology supported medicine. All of it, the lurking fears of infection, the saving graces of hygiene, the fears of ruining a child by affection, the selling of science, the desire of parents to learn from the experts, all came together to create one of the chilliest possible periods in childrearing. "Conscientious mothers often ask the doctor whether it is proper to fondle the baby," wrote an exasperated pediatrician in the late 1930s. "They have a vague feeling that it is wrong for babies to be mothered, loved, rocked and that it is their forlorn duty to raise their children in splendid isolation, 'untouched by human hands' so to speak and wrapped in cellophane like those boxes of crackers we purchase."

Oh, they were definitely saving children. In 1931, Brenneman reported that his hospital in Chicago was averaging about 30 percent mortality in the children's wards rather than 100 percent. Yet the youngest children, the most fragile, were still dying in the hospitals when they shouldn't. They were coming in to those spotlessly hygienic rooms and inexplicably fading away. At Children's Memorial, babies were dying seven times faster than the older children; they accounted for much of that stubborn 30 percent mortality. Brenneman also noted that babies who did best in the hospital were those who were "the nurses' pets," those who enjoyed a little extra cuddling, despite hospital rules. Sometimes the hospital could turn an illness around, he said, by asking a nurse to "mother" a child, just a little.

New York pediatrician Harry Bakwin had come up with a description for small children in hospital wards. He titled his paper on isolation procedures "Loneliness in Infants." French researchers had begun to suggest that the total "absence of mothering" might be a problem in hospitals. An Austrian psychologist, Katherine Wolf, had proposed that allowing a mother into a hospital ward could improve an infant's survival chances. She insisted that there might be actual risk from "the best equipped and most hygienic institutions, which succeeded in sterilizing the surroundings of the child from germs but which at the same time sterilized the child's psyche." Did this make sense? Absolutely—today. At the time, absolutely not.

Hadn't psychology declared that children didn't need affection and mothering? Why would anyone even consider the notion that hygiene and that wonderfully sterile environment might be dangerous to a child? The idea was just silly; so silly, so ridiculous, so trivial, in fact, that the field of psychology pretty much ignored Wolf, Bakwin, Brenneman, and the whole idea. Years later, British psychiatrist John Bowlby went hunting for studies of the relationship between maternal care and mental health. He could find only five papers from the 1920s in any European or American research journal. He could find only twenty-two from the 1930s. What he found instead were thousands of papers on troubled children—on delinquent children, children born out of wedlock, homeless children, neglected children. Neglect, as it turned out, bred neglect beautifully. As one physician wrote, "The baby who is neglected does in course of time adjust itself to the unfortunate environment. Such babies become good babies and progressively easier to neglect."

In a curious way, it took a war to change things, and a major one at that, the last great global conflict, World War II. Perhaps a minor skirmish would never have shaken psychology's confidence so well. It was an indirect effect of the war that actually started catching researchers' attention. Bomb fallout, the smashing apart of cities across Europe, the night bombings of cities by the Germans, the counter-bombings of the Allies, street after street in London blown apart, Dresden fire-bombed into a ruin of ashes: As the fires blazed, as their homes and streets shattered around them, many parents decided to protect their children by sending them away. They hustled their offspring out of the big-city targets to stay in the homes of friends or relatives or friendly volunteers in the countryside. In England alone, more than 700,000 children were sent away from home, unsure whether they would see their parents again. "History was making a tremendous experiment," wrote J. H. Van Den Berg, of the University of Leiden. It was impossible to deny the emotional effect on these children; they were safe, sheltered, cared for, disciplined—and completely heart-broken.

Austrian psychologist Katherine Wolf listed the symptoms: Children became listless, uninterested in their surroundings. They were even apathetic about hearing news from home. They became bed-wetters; they shook in the dark from nightmares and, in the day, they often seemed only half awake. Children wept for their parents and grieved for their missing families. In the night, when the darkness and the nightmares came calling, they didn't want just anyone; they wanted their mothers. Nothing in psychology had predicted this: Wolf was describing affluent, well-cared-for children living in friendly homes. It was startlingly clear that they could be clean and well fed and disease-free—you could invoke all the gods of cleanliness and it didn't matter—the children sickened, plagued by the kind of chronic infections doctors were used to seeing in hospital wards. It seemed that having good clean shelter really didn't always keep you healthy. The refugee children were defining home in a way that had nothing to do with science at all.

Bakwin, by that time, was blistering up the medical journals. He had supplemented the signs at Bellevue that said "Wash Your Hands Twice Before Entering This Ward" with new ones declaiming "Do not enter this nursery without picking up a baby." In a paper published at the height of the war, in 1944, he described hospitalized babies in a way that sounded startlingly like the separated children in England. The medical ward infant was still and quiet; he didn't eat; he didn't gain weight; he didn't smile or coo. Thin, pale, he was indeed the good baby, the easy-to-neglect baby. Even the breathing of these children was whisper-soft, Bakwin wrote, barely a sigh of sound. Some infants ran fevers that lasted for months. The simmering temperatures didn't respond to drugs or anything the doctors did. And the fevers, mysteriously, vanished when the children went home. A doctor ahead of his time—by a good three decades—Bakwin won support he needed from his superiors at Bellevue to let mothers stay with their children if it was an extended illness. He liked to point out that with the mother around, fatal infections had dropped from 30–35 percent to less than 10 percent in 1938, and this was before the availability of drugs and antibiotics became widespread.

"The mother, instead of being a hindrance, relieves the nurses of the care of one patient and she often helps out in the care of other babies." But Bakwin and Bellevue were an odd-island-out in the sea of medicine. Standard hospital policy in the 1940s restricted parents to no more than a one-hour-long visit a week, no matter how many months the child had been there. Textbooks on the care of newborns still rang with the voice of Luther Holt and the dread fear of pathogens. Experts continued to recommend only the most essential handling of infants and a policy of excluding visitors. Even in the 1970s, a survey of wards for premature infants found that only 30 percent of hospitals allowed parents even to visit their babies. And less than half of those hospitals would allow a parent to touch her child.

Bakwin argued that babies are emotional creatures, that they need emotional contact the way they need food. Of course, he put it in words becoming to the doctor he was: "It would appear that the physiologic components of the emotional process are essential for the physical well-being of the young infant." But he wasn't afraid to suggest that this could be a bigger problem than just what he saw in hospital wards. Orphanages and asylums also ran on the sterilization principle. And although children might stay days, weeks, occasionally months, in a medical ward, they might stay years in the foundling homes. Bakwin gave a simple example of the problem, centered on what might seem a trivial point: smiling. Somewhere between two and three months, he pointed out, most babies begin to smile back at their parents. "This is not the case in infants who have spent some time in institutions." They didn't return a smile. He and his nurses, if they had time, could coax a response, but there was nothing spontaneous about it and they often didn't have time. What if the child stayed longer? What would happen to her then? Or him? If people couldn't make you happy as a baby, could they ever?

Another New York physician, William Goldfarb, was also becoming worried about the fate of children in homes. The foundling homes were like a magnified version of a hospital ward; the emphasis was on cleanliness, order, self-control, discipline. Since psychology had declared affection unnecessary—perhaps even detrimental—to healthy child development, no one was wasting much warmth on these children, who were unwanted anyway. In the homes, youngsters were fed, clothed, worked, praised, punished, or ignored, but policy did not direct that they be cuddled or treated with affection. Often homes discouraged children from even making close friendships with each other because such relationships were time-consuming and troublesome. Goldfarb worked with Jewish Family Services, which operated a string of foster homes around the city. The children he treated were like the bomb escapees—apathetic, passive, and, which he found most troubling, they seemed to be extending their isolation zone. The foundlings often appeared incapable of friendship or even of caring about others. "The abnormal impoverishment in human relationships created a vacuum where there should have been the strongest motivation to normal growth," he wrote in 1943. At least children in their own homes—even if they had cruel or hostile parents—had some thread of a relationship that involved emotional interaction. The vacuum, Goldfarb insisted, was the worst thing you could inflict on the child, leaving a small boy or girl alone to rattle about in some empty bottle of a life. The younger they were thus isolated, the worse the effect. "A depriving institutional experience in infancy has an enduring harmful psychological effect on children," he said, and he meant all dimensions.

Two other New York–based researchers, David Levy and Loretta Bender, took up the cause as scientists in that urban community began sharing concerns. Like Bakwin, Loretta Bender worked at Bellevue; she headed the hospital's newly created child psychiatric unit, and many of her clients came from foundling homes. They were "completely confused about human relationships," she wrote; they were often lost in a fantasy world that might have served as a kind of shelter were the fantasies were not so ugly. The children spun their worlds hot with anger, cold with visions of death. If this was evidence of how foundling homes raised the youngsters, they were not producing anything that looked like normality.

Levy's interest began at another end of the spectrum. Starting in the late 1930s, he had decided to study those overprotective mothers so criticized by Watson. He wanted to compare extremes: thoroughly watched-over children versus motherless foundlings. He did find some very unhappy children held tight under domestic wings. Some were desperate for escape, some inhibited into near silence, some arrogant and exhibiting a sense of entitlement. The foundlings he met were often silent or desperate. But they were often unnerving, as well. Many of the orphans had learned starched and polite manners. Too often, Levy couldn't move past that polished amiability. Neither, it appeared, could anyone else. The foundlings, especially long-time ones, were the well-behaved strangers at a party who have perfect manners and complete inner indifference to you. Those upright behaviors did sometimes get them adopted. But they inevitably chilled the affection out of such relationships. One hopeful mother, after a year of trying to coax some warmth out of her adopted child, returned the little boy. She said that she felt that she had been punished enough. "Is it possible that there results a deficiency disease of the emotional life, comparable to a deficiency of vital nutritional elements within the developing organism?" Levy wondered.

Of course, this was a worry mostly still buried in academia, a matter of research journals and scientific debates. The lonely-child syndrome that Bakwin described so eloquently had a name: "hospitalism." But what did that mean? Most people had never seen a child suffering from hospitalism, or watched a baby spiral down in his weeks on the ward. Bakwin could write of the despairing sigh of a child's breath. He could draw a heart-wrenching portrait of the way a lonely baby would begin to wither, until he began to look like an old man. And Bakwin did do that, all of that, with determined eloquence. But his words, however frustrated and angry, were still words in a medical journal. They were read and debated by a select few. It seemed that to change the picture, some advocate of the lost child would need to think about a far wider audience.

Scientists like to work within their own community, communicate in their own jargon, publish in their own journals. But to be a crusader, one must sometimes push beyond the academic envelope. John Watson had understood that perfectly—and used it to remarkable effect. Researchers working with orphaned children were reaching that same awareness. They would need the power of public opinion to change the system. They would need to make people *see* the problem, literally. The power of the filmed image suddenly beckoned as a way to break through the refusal to find out what children needed. In particular, a Viennese psychiatrist named René Spitz and a Scottish medical researcher named James Robertson both came to that conclusion. Spitz and Robertson, on different continents and for different reasons, decided that words were never going to win this fight. Each one decided to find a movie camera. Each would attempt to show people exactly what was being done to children.

Spitz was a Vienna-born Jew who fled from Austria to France, and then from France to New York, as Hitler's armies spread across Europe. He had worked with Katharine Wolf in Austria on the issue of sterile children's wards. In New York, he settled down with a passion to join forces with the likes of Harry Bakwin and William Goldfarb. In 1945, he was the author of yet another research paper, "Hospitalism: An Inquiry Into the Genesis of Psychiatric Conditions in Early Childhood." If one reads beyond the scientific terminology, his paper tells the compelling story of four months that Spitz spent comparing two sets of children. None of the children was blessed in his circumstances. One group consisted of infants and toddlers left by their parents at a foundling home. The others attended a nursery school attached to a prison for women.

Spitz's description of the foundling home would have a familiar feel to anyone following Bakwin's work. The place was gloriously clean. Each child was kept in a crib walled off with hung sheets—or what Spitz tended to call "solitary confinement." The home observed the common practice of "don't touch" the child. Masked and gloved attendants bustled around, arranging meals and delivering medicine. Still, the only object the children saw for any length of time was the ceiling. In spite of "impeccable" guards against infection, the children constantly tumbled into illness. The home housed eighty-eight children, all less than three years old, when Spitz arrived. By the time he left, twenty-three were dead, killed by relentless infections.

The nursery, by contrast, was a chaotic, noisy play place, a big room scattered with toys. Children constantly tumbled over each other. The prison nursery allowed mothers to stay and play with their children. Perhaps because it was such a break from cell life, the mothers did as much as possible. Or perhaps they just wanted to be in a place where they found plenty of hugging and comfort. None of the children there died during Spitz's study. That didn't mean that you could blame all the deaths on loneliness. But, Spitz insisted, it should be considered as a legitimate peril, a recognized threat to health.

The "foundling home does not give the child a mother, or even a substitute mother," Spitz wrote. There was one staff attendant for every eight children, or what he called "only an eighth of a nurse." The problem with solitary confinement, he argued, is not that it's boring or static or lacks opportunities for cognitive stimulation, although all of that is true, and none of that is good. The more serious problem for the children was that there was no one to love them. Or like them. Or just smile and give them a careless hug. And it was this, Spitz said—isolation from human touch and affection—that was destroying the children's ability to fight infection. At the center of Spitz's argument is a simple statement: For a child, love is necessary for survival. His first choice to provide that was the mother. He wouldn't turn away others, though—an affectionate caretaker, a person actually interested in the child, someone more than one-eighth of a nurse. Any and all of those people were, he thought, a medical necessity: "We believe they [the children] suffer because their perceptual world is emptied of human partners," he said flatly. What is life without a partner? Can there be a home without someone who welcomes you there?

Spitz found that his paper received, well, mild interest, moderate attention. It added to the ongoing argument—the one that was going nowhere.

Spitz prepared to fight harder. He had filmed the children as they came into the foundling home and had allowed the camera to continue observing as the weeks passed. Simmering with his own outrage, Spitz turned his grainy little black-and-white film into a 1947 psychology classic, a cheap little silent movie, its title cards crammed with furiously compassionate words. He called the film simply, *Grief: A Peril in Infancy.* It starts with a fat baby named Jane, giggling at the experimenter, beaming at the people around her, reaching to be held. A week later, Jane sits in her crib, peering constantly around, searching for her mother. She is unsmiling and, when Spitz picks her up, she breaks into uncontrollable sobs; her eyes are pools of tears. There's the next little girl, "unusually precocious" says the title card, seven months old, happily stroking Spitz's face, shaking hands with him. A few weeks later, she's pale, unsmiling, dark circles curve under her eyes. She won't look up at Spitz now. He gently raises her from the crib. And then she clings to him so desperately that he has to pry her off when he leaves. She's still sobbing when the camera turns to another baby, lying flat, staring into the air, pressing a fist against his face; and another, curled up, trembling, gnawing on her fingers. The title card this time is short and indeed to the heart of the problem: "The cure: Give Mother Back to Baby."

Spitz took his film from medical society meeting to medical society meeting in New York. In his eloquent book on the importance of early relationships, *Becoming Attached*, psychologist Robert Karen writes that one prominent analyst marched up to Spitz with tears in his eyes, saying, "How could you do this to us?" The film did indeed cause the debate over mother-child relationships to steam. Could Spitz be right? Could some fifty years of psychiatry be so wrong? Even eight years after *Grief* was produced, the quarrel still simmered. Critics shredded the film all over again as emotionally overwrought and nonscientific. Even in the late 1960s, researchers were arguing over whether he was right. But it was almost impossible, as Spitz had known, to argue those weeping children away.

Another film was circulating by this time, James Robertson's documentary of children in medical care. It was a cheap little film, too. Robertson estimated that it cost $80 to produce. His was a different story from Spitz's—and the same. Robertson wanted to tackle children in hospital wards and what it cost them to feel abandoned by their parents. This was still, of course, during the time of brief weekly visits. He called his film *A Two-Year-Old Goes to the Hospital.*

For a child at that time, hospitalization was, essentially, isolation from home and family and friends and everything that might have given a sick child security and support. Robertson's film followed a poised little toddler named Laura. He said once that she was so naturally composed that he worried that her very temperament would render his case meaningless. And Laura did indeed go easily into her hospital bed. But by the next week, she was begging her parents to take her home; and the next, pleading with them to stay; and by the next, hardly responding to them at all, just her lips trembling as they left her behind. At the end of the film, she was like a frozen child, silent and unresponsive. Months later, Laura, back home and secure again, saw Robertson's film, turned to her mother, and said angrily, "Why did you leave me like that?"

Robertson showed his film to an audience of three hundred medical workers in England. The initial reaction was concentrated fury. The hospital staffers felt personally attacked. Many demanded that the film be banned. "I was immediately assailed for lack of integrity,"

Robertson recalled. "I had produced an untrue record. I had slandered the professions." In 1953, Robertson became a World Health Organization consultant and brought his film to the United States for a six-week tour. Here, again, he ran into a solid wall of defensiveness, as if the ghosts of John Watson and Luther Holt were rising up in revolt. Robertson was assured that the problems he had documented were British ones: "American children were less cosseted and better able to withstand separations." And his simple solution—let parents stay with their children—was rejected as wrong-headed.

Robertson, though, had an unusual ally who liked the film and the message behind it. Edward John Mostyn Bowlby, born in 1907, was the son of a baronet. His father had been surgeon to the royal family. The son had been raised in time-honored upper-class style—a nanny until he was eight and then off to boarding school. It hadn't been a happy experience. John Bowlby later told his wife that he wouldn't send a dog to boarding school. Bowlby's father had wanted his son to follow him as a physician. He obediently entered medical school at Cambridge, but finally rebelled against doing as he was bid. Bowlby dropped out of the university and spent two years working in schools for troubled children. That time, and the almost heroic struggles of children seeking some kind of balance, decided Bowlby on a career in psychiatry. In 1929, he entered medical school at University College Hospital to train as an analyst. In time, he would indeed become a smart and thoughtful psychoanalyst. He figures in this story, though, because he would also become more—a brilliant theoretician, a world-class crusader.

Psychoanalysis belonged to one man at the time, and that was Sigmund Freud. When Bowlby began training as a psychiatrist, Freud was seventy-three years old, living in an affluent section of Vienna. Within the next decade, the Nazis would confiscate Freud's home, his money, his publishing house, and his library, and kill all his sisters in the gas chambers. He, his wife, and his children escaped to England in 1938, but Freud never recovered. He died of cancer within a year of arrival on safe soil. Yet even in the last ragged years of his life,

Freud cast a long and powerful shadow. He still does, of course, more than sixty years after his death. In Bowlby's time, it was a living shadow, as if some smoky image of Freud were still standing by, frowning at one's mistakes and one's doubts about his theories. His daughter, Anna Freud, helped keep his influence alive. She became one of the dominant psychoanalysts in post–World War II Britain. But Freud's ideas stood on their own power. They were potent enough, provocative enough to continue challenging the field indefinitely. The years since Freud died have stayed full of his ideas—of the subconscious mind, of sexual repression, of the power of a fantasy life. The smoky figure has faded, but not away, ever, entirely.

The aspect of Freud's theories that Bowlby found so difficult had to do with reality. Freud had declared that the unconscious in the adult is "in large measure made up of the child slumbering within, the child who dreams and fantasizes of a better life, so intensely that sometimes the adult cannot distinguish the two." And neither, Freud suggested, could the child. In other words, a child might be most heavily affected by his fantasy life and not by real events. This would mean that what a parent might do to a child was not nearly as important as the child's internal perceptions and desires and fantasies about that parent. A mother's touch might be meant as affection, for instance, but be turned into sexual dreaming by the child. If a child reported sexual abuse, then, it might only be the manifestation of desire. Perhaps the memory of a seduction was actually the memory of a wish. A sexual dream woven out of equal parts imagination and longing. Young children, Freud said, have a potent erotic drive that causes them to *want sex* with their opposite-sex parents. Reality doesn't have to enter into it at all.

Freud didn't say that early connections were meaningless. Shortly before his death, he wrote that the tie with the mother was "unique, without parallel, laid down unalterably for a whole lifetime" as the prototype for all other relationships. On the other hand, he still said, that unparalleled relationship didn't have to be entirely *real.* The child might be influenced by his perceptions of something his mother had done, or his dreams of her, or even those lingering erotic fantasies. Spitz could argue that baby needed mother; Goldfarb could argue that children must learn affection when young; Bakwin might insist that babies are emotional creatures. But if doctors were looking for professional support in keeping mother and child physically together, they were not yet going to find it in the community of Freudian psychoanalysis. Anna Freud once explained it like this: "We do not deal with happenings in the real world but with their repercussions in the mind."

So when John Bowlby trained in psychiatry, he was startled to find that "it was regarded as almost outside the proper interest of an analyst to give systematic attention to a person's real experiences." It didn't take Bowlby long to realize that he couldn't work that way. His time with the maladjusted school children had convinced him of the power of real life. He knew that how parents treated children—if they had parents—mattered intensely. In 1948, working for the World Health Organization, Bowlby took his stand, beginning with a report titled *Maternal Care and Mental Health.* In it, he gathered together his allies. The report rings with the work of Bakwin, Goldfarb, Spitz, Bender, and other observations, including Bowlby's own.

Scientists who knew Bowlby remember him as almost a stereotype of the British gentleman, sometimes arrogant, dry in humor and tone, unsentimental, outwardly cool. But in the WHO report, he is passionate. Anger hums in the pages like electricity through a wire: "The mothering of a child is not something which can be arranged by roster; it is a live human relationship which alters the characters of both partners. The provision of a proper diet calls for more than calories and vitamins; we need to enjoy our food if it is to do us good. In the same way, the provision of mothering cannot be considered in terms of hours per day but only in terms of the enjoyment of each other's company which mother and child obtain."

Another concept, beloved by the Freudians, was that the baby's first relationship was not with the mother as a whole, but with her breast. Infants, so the thinking went, lacked the mental capacity to form a relationship with a whole person, or even to keep the concept of a person. When Freud wrote of mother love, he also explained that the breast that feeds is an infant's first erotic object, and that "love has its origin in attachment to the satisfied need for nourishment." Bowlby had studied under another dedicated Freudian psychoanalyst, Melanie Klein, who agreed that the most important "being" in an infant's life was the breast. The mammary relationship, so to speak, would define the child's connection to its mother. This was Freud's "oral stage" of development, the mixing of nourishment with a faint tinge of erotica. After World War II, when she had worked with displaced children, Anna Freud was more willing to discuss the notion that a child might love a mother. But she didn't believe that bond began in affection: "He forms an attachment to food—milk—and developing further from this point, to the person who feeds him and the love of the food becomes the basis of love for the mother."

This dovetailed beautifully with psychology's faith in the conditioned response—the baby is hungry, his hunger drive is satisfied, he becomes conditioned to associate his mother with food. Mother and breast are equal; good mother means good feeding. It was another perfect meeting of the minds in defining human behavior. There was Freud and his followers and their faith in fantasy and food. There was the conviction of mainstream psychology that affectionate mothering was irrelevant and that children could and should be trained. There was the medical profession's reluctance to believe that health and emotions were in any way connected. "It's hard to believe now," says psychologist Bill Mason of the University of California-Davis, now an expert in social relations, "but when I first started working in Harry Harlow's lab, the prevailing view in psychology was that a baby's relationship to the mother was based entirely on being fed by her."

By the late 1950s, despite the films and arguments and reports, the baby and the mother remained loveless in psychology. John Bowlby was running out of patience. He published another paper, or you could say another salvo, titled "The Nature of Child's Tie to His Mother" that was flatly grounded in the everyday reality of touch and affection. It was also his first attempt at putting forth his own theory of mother-child relationships, today known far and wide as attachment theory. And what attachment theory essentially says is that being loved matters—and, more than that, it matters who loves us and whom we love in return. It's not just a matter of the warm body holding the bottle; it's not object love at all; we love specific people and we need them to love us back. And in the case of the child's tie to the mother, it matters that the mother loves that baby and that the baby knows it. When you are a very small child, love needs to be as tangible as warm arms around you and as audible as the lull of a gentle voice at night.

Yes, Bowlby said, sure, food's important. But we don't build our relationships based on food. We don't love a person merely because she comes in carrying a bottle of milk or formula. We don't seek her out, clinging to her, sob when she leaves, just because she can feed us. That's lower in the hierarchy of needs—in the terminology of psychology—a secondary drive. Love is primary; attachment is primary. In Bowlby's view, a whole and healthy baby will want his parent nearby and will work for it—"many of the infant's and young child's instinctual responses are to ensure proximity to the adult." Babies aren't stupid; they know who will watch over them best. In attachment theory, a plethora of the infant's behaviors target mom or dad: sucking, clinging, following, crying, and smiling—perhaps cooing and babbling as well—are all part of the instinctive way a child tries to bind his parent tight.

There's a Darwinian side to this, Bowlby said, because a nearby parent undoubtedly increases the survival chances of the offspring. Without these behaviors, if parents lost interest, "the child would die, especially the child that was born on the primitive savannas where people first evolved." And, yes, obviously, food is necessary to survival, but it's a byproduct of the relationship. A baby knows that if the mother is there, she will provide food. Equally important,

Bowlby said, if the mother isn't there, not only is there no food but no protection against predators, and cold, and all the dangers of the night. So you might logically expect that we would evolve to be afraid and even despairing if our parents suddenly disappear. If you see a baby who appears to be suffering in his loneliness, Bowlby said, then you are seeing reality.

Push a child away, abandon it, and you do not see a well-disciplined miniature adult. You see the sobbing child in Spitz's film; James Robertson's Laura, clinging to her parents' hands; Bakwin's grave and shrunken babies in their screened-off beds. Bowlby's studies showed that, as children grew older, became toddlers, this need didn't lessen at all. The older children were just more aware. They knew their mothers better. They grieved when their mothers left them. They mourned a loss. They wanted their mothers back. In Bowlby's theory, this was a natural childhood reaction, like fear of the dark, of loud noises, strange people, and shadowy forests. If a baby's call wasn't answered, the child was left to fend for herself, make her own defenses. This could be part of what Goldfarb saw in the emotionally cold children from orphanages. Their emotional distance might be self-protective, Bowlby agreed, because it buffered away grief and loss. But it could also be destructive because "it sealed off the personality not only from despair but from love and other emotions."

Bowlby's ideas angered almost everyone he knew. Anna Freud dismissed him outright. She sincerely doubted that infants had enough "ego development" to grieve. Klein accepted that an infant might look sad, go through a "depressive" stage; but that wasn't missing a mother, she said, that was normal development. All Bowlby was seeing, she insisted, was reaction to sexual tensions, probably just baby castration fears and rage against dominating parents. The British Psychoanalytic Society was so hostile to attachment theory and its author that Bowlby stopped going to the meetings. "Unread, uncited, and unseen, he became the non-person of psychoanalysis," wrote Karen.

For the moment, all that compassionate momentum on behalf of children seemed to have stalled. It was beginning to look like a noble but lost cause. Perhaps that's exactly what attracted Harry Harlow to the research. That's not to say that the call was immediate. When Harry graduated from Stanford, John Watson still ruled, and there was no one around to take young Professor Harlow particularly seriously. Stanford hadn't; and, as it turned out, when he arrived in Wisconsin, his new university didn't, either. To hoist a banner in the name of love, Harry Harlow was going to need more than a name change. He would have to persuade other psychologists to listen to him. He would have to prove that his opinions mattered. He would pursue those goals in the least predictable ways: conduct experiments at a zoo, hand-build a laboratory, become obsessed with the intelligence of monkeys, and become convinced that he could, and should, quarrel with his own profession. You could call it an unusual route to the advocacy of love and affection. But there was never anything conventional about Harry Harlow.

THREE

The Alpha Male

We speak of love, but what do we know about it, unless we see the power of love manifested; unless we are given the power to bestow and a willing heart to bestow it on?

Inscribed on the northeast wall of Memorial Church, Stanford University

THERE ARE OBVIOUS PHYSICAL DIFFERENCES between Stanford and the University of Wisconsin, starting with water. The Madison campus overlooks a tree-rimmed lake rather than the sharp edge of the Pacific, a vista pretty rather than breathtaking. In the summer, Lake Mendota dances with wind-ruffled wavelets of light. In the winter, the waves freeze solid and unusually fanatical fishermen venture out on the rough gray-green surface and drill through to the frigid waters below to drop their lines. The campus, rambling above water level, changes with the lake. The tree-dense hills blaze like flame in the fall, turn white as ash in the long, long winters. The inevitable snow and ice and the frozen wind off the lake produced in Harry nostalgic memories of the same season at Stanford. Even the old slights and insults could take on a golden tint of warmth: "They expected to place me in a California junior college," he once said, "and with every Wisconsin winter, I wish to God they had."

But it wasn't the ice-rimmed winds or the sudden shift from graceful Italian architecture to sturdy sandstone that provided the real culture shock. It was the shift from Stanford's high-intensity program to Wisconsin's more easygoing approach. When Harry arrived in Madison, the psychology department had four faculty members, took on about three Ph.D. students a year, and was compact enough to be tucked into a basement of the administration building. It's hard to maintain visions of being an influential psychologist when no one can find you. In fact, Harry himself couldn't find the psychology department when he arrived on campus.

The university didn't cater to junior faculty. In case he'd missed that point, he had no map, no guide, and only the name of the administration building to get him there. "Excuse me," he said to two passing students. "Can you tell me where Bascom Hall is?" They looked at him. He was not quite twenty-five years old. Short, slight, with a rounded youthful face and curly dark hair. "Sorry," one of the students replied. "We can't help you. We're freshman too." Fortunately for Harry, Bascom Hall had presence even if he didn't. It loomed over the campus. The administration building sat atop the university's steepest hill, overlooking a new lawn, remnant forest, and lakefront. The multistoried hall was built of local pale gold sandstone and fronted with massive white Grecian-style columns.

Harry climbed the granite steps and, just inside the front doors, he found a reassuring sign on the wall. It was a black-and-white building directory listing "Harlow, H. F., Room 14." He wound his way into the basement, found his little cubby of a room, and sank down behind the desk, "savoring the first thrill of being a professor." Almost immediately, the door burst open again and a young man with a shock of wild dark hair stuck his head through the opening. He regarded Harry with dismay.

"Don't tell me he isn't here yet," the student exclaimed. "I absolutely must get started and I've been waiting to see him to know what to do. Have you any idea when this new man Harlow is coming?"

"Yes," Harry said.

It was apparent to him that being taken seriously at Wisconsin was going to be a lot harder than he had expected. In fact, learning to be taken seriously at Wisconsin was going to teach him just about everything he would need to know to be taken seriously elsewhere. He had an inkling of that on the opening day of his first undergraduate psychology class, an experience commonly described in his department as "being thrown to the wolves."

On the first day of class, Harry stood up in front of four hundred–plus freshmen and sophomores and was abruptly overwhelmed by his childhood shyness. His tongue tied. His r's disappeared. He tried mumbling them, but no matter. They sounded like w's. When Harry attempted to say "right" and it came out "wight," some of the students booed. "The first ones weren't very loud, but the next ones were," he recalled. By the end of class, he could hardly be heard over the catcalls and laughter. Not that he wanted to say anything else. He just wanted to get out of there.

Later, he would call that class one of his most important learning experiences. At the time, he was worried and hurt. In the early twenty-first century, when we work so hard to be tolerant of differences, we forget how culturally accepted intolerance once was. Thus Terman's almost unchallenged division of the world into the deserving gifted class and the undeserving stupid class. People scoring below the curve on Terman's Stanford-Binet were called "feeble-minded," remember, and that was one of the politer terms. They were also "mental defectives" or "morons" or even "undesirables." People who limped were "gimps." The homeless were bums. People who couldn't talk were dumb. And the Elmer Fudds of the world, people like Harry who struggled with an "r" here and there—they were ridiculous. It was standard procedure for students to boo a teacher if they considered him a joke.

At first, the situation seemed impossible. The more nervous Harry became, the more he stumbled over the dreaded r's. He decided to try appeasing the wolves in another way. He hunted up funny little anecdotes and jokes, relying heavily on *Reader's Digest,* and he started slipping them into his lectures. The students laughed at the stories but they still booed his pronunciation. Finally, he decided to give the students something to really boo about—a groaner of a joke, a truly appalling pun. Harry was a pun addict anyway; word play was like child's play to him, pure fun. He punned and the students groaned. He added a few more puns. As the puns became more obnoxious, the groans turned into boos. So Harry became even more outrageous. The boos grew louder. Increasingly, though, it was the jokes the students were reacting to. As Harry relaxed, he was stumbling over "r" less, anyway. He discovered, too, that if he slowed down the pace of his speech, kept an even rhythm, he could almost make those errant w's disappear. He developed such a clear, distinctive speaking style that a fellow psychologist once described Harry's voice "as cool and crisp as chilled lettuce."

Eventually, puns would come as naturally to Harry Harlow as poetry and breathing. When the university billed him for distilled water, he put up a notice on his bulletin board: "Distilled waters run steep." When a student asked him why male animals did better in certain tests, he snapped back, "They have to meet a stiffer criterion." Harry never completely conquered the r's, but eventually he learned not to care. He even credited the undergraduate wolves for confounding Terman's predictions and turning him into a speaker of national caliber. "Teaching elementary psychology. It's the best possible speech and timidity therapy you can have," Harry said.

He also credited someone else. He had an accomplice in figuring out how to thwart the wolves, a graduate student in psychology, Clara Mears, who rapidly became more than a friend. Clara was the daughter of a Congregationalist minister. She had little tolerance for cruelty and was delighted when Harry won the teaching war. "The only trouble," she liked to say, "was that he never did stop punning." And that made Harry laugh. The two of them seemed a natural support system. Friends and colleagues and family encouraged the relationship. So did Harry's old professor, Lewis Terman. Harry's friendship with Clara Ernestine Mears had an odd small-world twist to it.

She wasn't just any bright graduate student. She was a charter member of Terman's study of gifted children.

Clara was small, warm, and exuberant, "like a pet kitten," her mother said. "One did not readily associate sorrow with her." She had big brown eyes in a round face, a quick crisp voice, and a chuckle like a tumbling brook. She was born July 8, 1909, in Reno, Nevada, and was fourteen when both she and her older brother, Leon, were recruited into the gifted project. Clara was the youngest of five children in the Mears family and ever the most confident. Her mother described her on Terman's questionnaire as follows: Clara started reading at age four. She progressed to poetry at age five and her favorite author that year was a seventeenth-century poet, William Blake. By age eleven, she still liked British poets but had moved into the nineteenth century and now preferred Robert Browning. Reading was her favorite hobby, period. But her talents weren't limited to the literary. By the time she was a Browning fan, she was also doing her older sisters' algebra work for them.

Her mother's only complaint was that Clara was so indifferent to domestic chores. "She reasons someone else into it if possible. She says she will live by her brains instead of handwork." Clara liked to cook if it was creative enough. But housework, mending, the average Sunday dinner? It bored her bright daughter right out of the house. "She plans to hire someone to do that," Ernestine Mears wrote mournfully.

The ambitious scholar graduated from high school at the age of fifteen. She went on to a private women's school in the San Francisco bay area, Mill's College. There Clara began to appreciate her own abilities. "Dearests," she wrote to her parents in 1928, to tell them that she had outscored her entire class on an aptitude test. "Isn't that quite astonishing? Of course, from the natural run of events I've always known I could go faster than average, but I didn't know quite where I was." She decided to go on to graduate school. In 1930—the same year that Harry joined the faculty—the University of Wisconsin offered her a research assistantship in psychology. One of her first classes was Harry Harlow's graduate seminar on emotions. She liked it—or perhaps him—enough that she signed up for his next course, an evening seminar in physiological psychology. "When I began making A's in physiological psychology, Harry Harlow began escorting me home," she said.

Clara's outgoing friendliness coaxed Harry out of his natural shyness. They went to parties together, shared dinners. They played tennis and bridge and squash with fierce competitiveness. Clara, even more than Harry, loathed losing. "She does not like to be beaten in games but Harry is laughing her out of that," her mother wrote. And Clara, herself, made a happy confession to Terman. With Harry, she felt she could just be herself. He was a man "who admires accomplishment but never demands it or suggests or overrates it instead of a father whose influence was always toward the peak." With Harry, she had a sudden rush of pleasure in being liked just as she was. Her parents saw a kind of visible happiness. "There's a peculiar gleam or shine or radiance from her eyes. Our oldest girl was a beauty, but Clara exceeds her by now by far," Mrs. Mears wrote happily to Terman.

Harry Harlow and Clara Mears married May 7, 1932, in Milwaukee. Terman fired off a congratulatory letter, telling the newlyweds that he rejoiced in the marriage of Clara's "splendid hereditary equipment" to "one of the most productive young psychologists in America." In fact, married life would have started off as almost pure celebration if not for the University of Wisconsin.

Like most institutions at that time, the university had an inflexible nepotism policy. It did not allow spousal hires. As written, this policy might sound neutral. In practice, it wasn't. In the 1930s, the faculty was almost entirely male. This meant that it was usually wives, not husbands, who were kept out. It didn't matter that Clara was a promising psychologist, that even the famed Lewis Terman thought her exceptionally smart. Clara's advisor recommended that she drop out of her Ph.D. program. It would be a waste of time to continue, she was told, because Harry would always be the first choice for a job in psychology. Many years later, Clara would still bubble with resentment. She lost her career in psychology and, eventually, she lost her sense that she and Harry were intellectual equals. She thought Harry felt the same. It would take her a long time to regain faith in herself—or in their relationship.

But at the time? She was happy anyway; they were in love, and she was resigned. She took a job as a sales clerk in the dress department of one of Madison's department stores, an elegant, locally owned company called Harry F. Manchester. Here, indeed, Clara lived up to Terman's assessment of her abilities. Within six months, she was the store's chief dress buyer. She liked stylish clothes and enjoyed telling people how to dress. She liked the salary; it allowed her to be generous. She sent some of her paycheck home to her parents every month. Retired ministers, it seemed, didn't always have the spending money that dress buyers did. On one of Terman's surveys, Clara said that her favorite personal quality was "a sense of humor. . . . I can't take troubles too seriously but I do face them." She was determined to make the best of her unexpected career change. She studied fashions, made herself a model of perfect grooming (she described her least favorite personal attribute as her own relentless perfectionism).

Still, Clara wrote cheerfully to Terman, "There were few good academic possibilities with my husband a professor. Now I find business more exciting, with the varied contacts and demands." And Terman wrote back, perfectly reflecting the attitudes toward working women at the time: "I thoroughly approve of such careers for wives. Only I hope some time that you can take a vacation and start a family. You see I am already looking forward to the enlargement of our gifted group."

For Harry, the dilemma was not giving up a career in psychology; the dilemma was getting one started. The university had provided him with an office. The department had cast him into the undergraduate teaching pit. It had not provided him with a research laboratory or any kind of funding for research. At Stanford, he had been told there would be a rat laboratory. But when he asked about the facilities, his department head told him that they'd decided to tear the lab down. There was no plan—whatsoever—to replace it. He was stranded. He was an experimental psychologist with no way to conduct experiments. He was a researcher with nothing to study. He was an animal psychologist without rats; at that moment he could be compared to an astronomer without a telescope, a marine biologist with only a jar of distilled water to study.

Later in his life, Harry enjoyed making wisecracks about studying rats, or, as he liked to call it, rodentology. He made lots of them. At the time, though, he saw nothing funny in the vanished rat lab. Rat research was all he knew. Rats were the gold standard of behavioral research. There was the occasional rabbit, guinea pig, dog, or cat, sure. For credible experiments, though, the profession had adopted the rat, the rat, and nothing but the rat. In fact, by some standards of the day, if you couldn't study rat behavior, you had little to say about human behavior.

By 1939, when Harvard psychologist Gordon Allport—a stately, white-haired radical if there ever was one—became president of the American Psychological Association, rats simply ruled. Allport recalled being approached by a passionate behaviorist, a follower of John B. Watson, who challenged him to name one psychological question that could not be solved by using rats as subjects. Allport described himself as so taken aback that it took him a moment to think of one. Finally, and even then tentatively, he suggested "reading disabilities?"

Allport acknowledged that rat research had "delightful suitability" for the practice of objective science; yet, he argued, objectivity didn't always mean reality. He wasn't convinced that you could treat human behavior with the "sterilized" approach of physical science. People were complicated and reliance on rats tended to make them look simple: "I thought of how men build clavichords and cathedrals, how they write books and how they laugh uproariously at Mickey Mouse," Allport said in his presidential address. "Could the study of rats explain this?"

It's easy, from our vantage point, to wonder how psychologists could believe that rat behavior completely explained human behavior. Or vice versa. The answer belongs to the same process that led the scientists to argue that love wasn't credible. Both come, at least partly, from that overwhelming, early twentieth-century drive to make psychology a *real* science. In their efforts to purify psychology—to give it what Allport called the sterility of physics—leaders in the field tended to dismiss behavior that couldn't be measured and quantified.

How do you measure love in a rat? If psychologists were to strictly control conditions, they needed lab animals that they could manipulate and test in precise designs. They might want many animals or few. Blinded animals compared to those with sight. Infants compared to adults. Rats could be acquired in almost infinite supply, their every hesitation timed, their every reaction checked, their every heartbeat counted. The problem, of course, is interpreting the beat of a heart. Allport's contention was that his colleagues avoided that problem. They chose instead to suggest that all behaviors—human or rodent—could be simplified to stimulus-response situations. That meant that the rat could be a strong model for humans across most conditions—barring, perhaps, reading disabilities. In Watson's words: "The behaviorist, in his efforts to get a unitary scheme of animal response, recognizes no dividing line between man and brute."

Even today, with all the aids and artistry of neurochemistry, molecular biology, and imaging technology, scientists will still refer to the brain as a black box. Beautifully complex, amazingly flexible, sometimes transparent, sometimes completely opaque. In the early twentieth century, without such technical wizardry, the box seemed filled with impenetrable darkness.

Watson worried that he couldn't find even four psychologists who could agree on what an internal sensation *was*—much less the mechanisms behind it. The founder of the *Journal of American Psychology,* Clark University's G. Stanley Hall, printed a tirade on the amateurish nature of psychology in the late nineteenth century. Hall, who would later become Lewis Terman's major professor, wanted more of the precise science that was exemplified—he thought—by his own work on raising children. The papers submitted to his journal just exasperated him. Far too many were purported studies of spirit life, dream signs, and prophecies that were "thought by their authors to be psychological." Instead, he complained, the papers were "utterly uncritical" and unworthy. "The *Journal* can print only the most exact and scientifically important research."

By the time John Watson became a professor at Johns Hopkins University, in 1909, he had come to fear that American psychology was becoming a profession of scientific speculators who made educated guesses about that black box but generated no real data. Watson wanted hard facts and testable hypotheses. He turned, in desperation, to Russian psychology. The Russians, also frustrated by the unyielding brain, had decided to make it simple. One of the hottest books in Russian psychology during the late nineteenth century was titled *Reflexes of the Brain.* Its author, I. M. Sechenev, proposed that thought itself might be part of a physical reflex and that the brain was no more than a twitchy, responsive muscle. Sechenev's pragmatic approach to intellect fostered an even more pragmatic generation of young Russian researchers. The most famous of those was a lapsed seminary student named Ivan Petrovich Pavlov.

Pavlov loved research. He once wrote an essay for his students telling them that "science demands of a man his whole life. And, even if you could have two lives, they would not be sufficient." He had begun his career with a meticulous exploration of digestion, illuminating the mechanics so clearly that he would receive a 1904 Nobel Prize for those studies. His best-known work followed logically from there. His digestion studies were performed in dogs. Pavlov noticed that his animals drooled not only when they were near food but also when they heard the footsteps of approaching lab technicians. Now why would they salivate for the dull thud of shoe against floor? Could it be one of those psychical reflexes?

Eventually, Pavlov developed the science of the "conditioned response." The dogs had learned to associate the footsteps with food. He spent the rest of his life experimenting with such conditioned behaviors. To Watson's real admiration, Pavlov refused to speculate about what might be happening in the dogs' brains. He considered that an internal process, one he could not measure. Pavlov would not even allow that the dogs might "recognize" the sound of footsteps. A brain activity such as recognition, Pavlov insisted, was untestable and therefore "needlessly speculative." He had proved that dogs could be conditioned to drool in response to sound. That meant the brain could be trained—no more, no less. This methodical observation-based science was exactly what John B. Watson wanted in American psychology. When he thought of Pavlov, he said, he was determined to "give the master his due."

In his most extreme articulation of this philosophy, Watson proposed, in 1913, that perhaps the larynx controlled thought. Scientists could measure its activity. In those seconds or minutes (depending on intelligence) before people talked, the larynx was busily vibrating away. He reminded his colleagues that people don't seem as smart and articulate when they have sore throats. In a letter to a friend at the University of Chicago, Watson closed: "No matter where you are, or what you do, I'll be having a laryngeal perturbation about you every so often."

People are simple, Watson insisted, animals are simple. Dogs drool, we drool; rats develop conditioned habits, we develop the same. There are no higher humans and lower animals; he said that all are basic, all are regulated and driven by stimulus and response. By that logic, rats were an easy substitute for humans. And Watson preferred them as research subjects. But after World War I—in a curious parallel to Harry Harlow's dilemma—he found that his university had closed his rat laboratory. As he explained to a friend, human infants were probably the next best subjects for simple experiments in psychology.

If Pavlov could induce dogs to drool when they heard a metronome, Watson should be able to induce a baby to respond on cue as well. Watson began by simple conditioning of emotions, such as fear. He showed that you can make a baby afraid by sudden movement, an abrupt loss of support—such as by jerking a blanket out from under it. You can induce rage by pinning its arms down. You can condition love by stroking, he said. But could you condition a baby like one of Pavlov's dogs, train it to respond to an unexpected stimulus? The resulting study, published in 1920, is still referred to as the "Little Albert" experiment.

Albert was a fat little baby at the time, nine months old, as calm and steady in his reactions as a seasoned soldier. He didn't even get upset when a blanket was yanked away. Watson paraded strange objects before Albert: a friendly white rat, an energetic rabbit, a dog, a monkey wearing a mask, a monkey not wearing a mask, a wild fluff of cotton wool, even a burning newspaper. The child merely gurgled and watched with interest. He tried to pet the animals and fan the flames.

Loud noises, however, terrified him. The experimenters discovered that if they banged a hammer on a steel bar—clanging the metal close to his ear—the child jumped. His arms flew up in the air. If they continued to clang, Albert began to sob. Watson decided to see whether he could "condition" Albert to fear a harmless object, such as the rat, by producing a clang of metal every time the animal approached. Watson admitted to some ethical hesitation. Should a researcher deliberately and repeatedly scare a small child? But he reassured himself that life would naturally get tough anyway "as soon as the child left the sheltered environment of the nursery for the rough and tumble of the home."

When cheerful Little Albert was eleven months old, the scientists seated him on a table covered by a soft mattress. Then they showed him another rat. For a brief moment, his face lit. He held out a hand to pet the furry head and—bang—the metal rods clanged by his head. The child fell forward crying, burying his face in the mattress. They did it again: The rat appeared and with it the wham of metal against metal. And again. And again, until finally, whenever Albert saw a rat, he began to whimper. "The instant the rat was shown, the baby began to cry. Almost instantly, he turned sharply to the left, fell over on one side, raised himself on all fours and began to crawl away so rapidly that he was caught with difficulty before reaching the edge of the table." A year after the experiment, Albert still sobbed if they even showed him a fur pelt.

Little Albert underlined all the points that Watson wanted to make—that there was no essential difference between people and animals, that we could explain all behavior as simple conditioned reflex—at least all behavior worth explaining. Watson didn't do any further baby studies. Those who followed his behaviorist credo didn't do many, either. They didn't need to—the baby was now a proven proxy for rodents, and vice versa. Later, people would wonder whether Albert feared the rat or perhaps John Watson, who appeared on film to be handling children pretty roughly. Later, critics would also point out that one set of experiments with one child shouldn't really be considered proof of anything. A few people just didn't think children should be treated so by scientists. Watson was dismissive on a grand scale. "Society is in the habit of seeing them [children] starve by hundreds, of seeing them grow up in dives and slums, without getting particularly wrought up about it. But let the hardy behaviorist attempt an experimental study of the infant or even begin systematic observation of it and criticism begins at once."

You could definitely count Harry as a John Watson critic, given as he was to talking about "the Watsonian scourge." He thought the Little Albert study far too influential, describing it as "a trickle of data" that produced a flood of theory. He thought Watson's larger pronouncements about children were wrong to the point of being dangerous. He worried about all those children raised in the bleak Watsonian landscape: "For a generation there was no major mental institution in the country without its population of Watson-raised babies." Nor did he believe that all human behavior—Little Albert or no—could be compared to rats: "I am not for one moment disparaging the value of the rat as a subject for psychological investigation; there is very little wrong with the rat that cannot be overcome by the education of the experimenters."

Of course, given the value of rat studies when he first started in psychology, Harry might have become just such a researcher if the University of Wisconsin had given him that rat lab. He wanted to study the hot animal of the time, and he did his best to make that happen in a series of makeshift facilities. If they had worked, he might yet have stuck out rodentology. The medical school offered him one room for research, but it was so small that he kept tripping over the rat cages. He was then given two rooms in the attic of another building, where he stayed until summer arrived. At that point, "the near solar temperature" in the attics, Harry said, threatened to fry the brains of rats and experimenters alike. He next tried putting the rat cages into two small rooms, near his office, in the administration building's basement.

Bascom Hall was a dignified and solid structure. At least that was true of the administrators' offices on the upper levels. The windowless basement home for psychology was a poorly ventilated space partitioned into box-like rooms. Harry's particular box turned out to be directly below the Dean of Men's office. The pungent smell of rat bedding was sucked upward. Or, as Harry was rapidly informed by the administration, "noxious rodent odors floated fragrantly to the floor above." Nervous students, waiting for a conference with the dean, could now be found leaning out the windows. This last fact appealed a little, okay, a lot, to Harry's sense of humor.

In later years, when he was firmly entrenched as a primate researcher, Harry tended to think of his early research efforts as pretty funny stuff. Not just getting his rats kicked out of Bascom Hall, although he told that story to all his friends. He also studied cats in the basement of a fraternity house next to his apartment. That experiment involved putting a cat in a little cradle wired to give a weak shock. Harry and an assisting student would ring a bell and the cat would get that little tingle of electricity. After a while, the cats would jump out of the cradle if they just heard the bell. It was a classic study in Pavlov's techniques. "This process is called a conditioned response," Harry said. "One of our cats was a little sensitive. When we rang the bell, he jumped out of the cradle, out of the basement window and kept right on going." The professor and his student pelted up the stairs and out the front door of the frat house. They ran down the street, calling for the animal, and becoming increasingly winded. But, "it was a wonderful conditioned response. At least it was good for over a mile and a half, when we lost sight of the cat."

After his cat research, Harry tried similar experiments with frogs. Perhaps, he thought, amphibians could also be conditioned to respond to a ringing bell or flashing light. And undoubtedly they would be easier to catch. So Harry flashed lights. He rang bells. He applied mild shocks to the frogs' legs. The frogs never seemed to make the association. A shock was a shock and a bell was a bell—that was apparently how his amphibians saw the situation.

Harry was so exasperated by his frog experiments that one day he began venting to anyone who would listen. He even told his undergraduate class that he had spent countless hours just to prove that frogs were stupid. One of the students happened to be a reporter for the student newspaper, the *Daily Cardinal,* and the next day, Harry was in print: "Professor Harlow says that the frog is the dumbest of all animals. . . . Professor Harlow's experiments showed that the frog does not seem to be able to learn anything at all." It was a natural story for any journalist. The next day, one of the local papers rewrote the student version. It now carried the headline: "Frog Dumbest of Animals, Experimenters Discover."

That story was picked up by the wire services. Those wire stories led to a round of editorials on how scientists waste the taxpayers' money and whether frogs might actually be smarter than scientists. "I never knew why but this choice little bit of information was circulated far and wide," Harry wrote in his memoir. "My relatives scattered throughout the nation were amazed at my sudden and undignified rise to fame. My colleagues were amused or sarcastic. A couple of them suggested that I ought to see a psychiatrist."

Maybe therapy was the answer. But, finally, in sheer desperation, he went instead to visit the monkeys at the local zoo. The suggestion was actually made at a dinner party when Harry and Clara were partnering each other in bridge after the meal. Harry's conversation, just then, tended to focus on his nonexistent laboratory. Obsessively. Oh, said an opposing bridge player, but you should check out the orangutans at the local zoo. Never mind cats and frogs and rats, what you should study is a really interesting primate. As Harry later remembered, his fellow bridge player told him to stop worrying so much. All would be fine. She thought that orangutans had a lot more charm than rats—and more than people, too.

The Henry Vilas Zoological Park occupies a shady little pocket of land in the city of Madison, backed against a small lake called Wingra. The zoo grounds are designed as a friendly ramble amidst trees and cages. Picture for a minute Harry Harlow, in his late twenties, his face still so boyish that students continued to mistake him for a fellow undergraduate. He's standing moodily in front of the orangutan cage, shoulders slumped, hands wedged in his pockets, staring. The orangutans are staring back—apes do that, they're interested, too—and he's thinking, he's thinking . . . he's thinking, what the hell am I doing here? He's thinking, what else is there? And finally, he's thinking that at least one of the orangutans does seem rather charming.

The easy going ape, named Jiggs, shared his cage with an outstandingly irritable female named Maggie. The two orangutans had been named after a popular comic strip of the time. The strip was a running domestic comedy. The story line featured an easygoing doormat of a man named Jiggs and his hot-tempered, rolling-pin-waving wife, Maggie. The old orangutans might have modeled for the comic strip. Harry described Jiggs as "the nicest and sweetest orangutan that had ever lived at any zoo for fifteen years." Maggie, on the other hand, was "a girl with a one-track mind, a determination to keep Jiggs on the straight and narrow." Whenever the old male displeased her—which as far as Harry could tell was every few minutes—she slapped him. Standing on the pathway, the young psychologist winced in sympathy. He asked the zoo director, Fred Winkleman, whether it would be possible for Jiggs to be separated from his mate, at least long enough for Harry to try him out on some standard intelligence tests. Perhaps Winkleman was also sympathetic. He said yes.

So Harry Harlow and his startled students started making the one-mile trek from campus to the Vilas Zoo. They brought tools for testing—tables and trays and blocks and puzzles. Among the challenges they gave Jiggs was one familiar to many nursery school children: putting the right peg into the right hole. The puzzle consisted of two oak blocks, one with a square hole and one with a round hole, along with two plungers, one square and one round.

Jiggs loved it. He quickly learned to put the round plunger into the round hole. He discovered that it would also wedge into the square hole. He learned to put the square plunger in the square hole. But he was baffled by the discovery that the square peg would not go in the round hole. He worked on that problem, on and off, for hours. To Harry's genuine sorrow, the old orangutan died about a year after the testing started. "At least," Harry commented, "Jiggs died demonstrating a level of intellectual curiosity greater than that of many University of Wisconsin students."

Harry and his students went on to test a big male baboon named Tommy, who wasn't nearly as sweet in temperament. Tommy liked to get it right, and right away. When he made a mistake, he was furious. The researchers had built a testing table for the experiments. They would hide food under cups on the table. Tommy's task was to remember which cup hid the food. He was fine if he could look instantly. But when they were measuring length of memory, they would stave him off. In psychology, this is called a "delayed response trial." To Tommy, it meant that he had to wait to look for the food, which he hated. Tommy was an unusually big baboon, weighing close to ninety pounds. Even medium-sized baboons can have memorable tantrums. Tommy would throw the cups and grab the table, smashing it against the bars of his cage. The tests and the tantrums continued. Neither Professor Harlow nor Tommy was willing to back down. Cups continued to fly.

And then the monkey developed a crush on one of the female students. At least that was Harry's interpretation: "It was impossible to look at Betty and not know immediately that she was a girl," he explained. You could also reason that Betty was just plain nice to Tommy. She fed him grapes. He would reach out his arms and she would rub his hands and wrists. She let him groom her arm—a standard gesture of friendship among many monkeys. When Betty ran the tests and Tommy had to wait for his turn, he would shake the bars of the cage, he would beat the floor, but he didn't break the equipment anymore. Tommy thus passed the tests with great style. And Harry Harlow became hooked on primate research.

In Jiggs's determination, in Tommy's eagerness to please a friend, even in Maggie's bossiness, Harry saw behaviors far more complex, far more interesting, than popular psychology suggested. He saw personality and relationships. It was at the Henry Vilas Zoo, in the companionship of an irritable baboon and a good-natured orangutan, that he finally turned away from rat research. "These are not just monkey stories," he said. "They are human interest stories and the reason why the monkeys, as far as our research was concerned, were in and the rats were out. The monkeys were so very, very much like people. No rat would have fallen in love with Betty." He sometimes thought about how much those first days at the zoo had shaped his interests. And he also wondered about the direction he would have taken had the expected rat lab been provided. Harry doubted that it would have led him to study mother love. "In my fondest fantasies, I cannot envision a rat surrogate mother." Spending his days at the local zoo, Harry was starting to feel a lot less foolish. Indeed, he was beginning to feel downright lucky.

He was lucky in his first graduate student, too. He was assigned a passionate, Brooklyn-born independent thinker named Abraham Maslow. Professor and student bonded over a shared skepticism about the current trends in behavioral psychology. "Behaviorism has done a lot," Maslow wrote in his own journal. "It was the beautiful program of Watson that brought me into psychology. But its fatal flaw is that it's good for the lab and in the lab, but you put it on and take it off like a lab coat. It's useless at home with your kids and wife and friends. . . . If you try to treat your children at home in the same way you treat your animals in the lab, your wife will scratch your eyes out." Well, Maslow thought his wife would.

Maslow cared most about human behavior. He believed that the mission of psychology was to help people reach their best potential. He never doubted that people *had* a best potential—rich with love, kindness, compassion. "People are all decent underneath," Maslow wrote in his journal shortly after receiving his Ph.D. from Wisconsin. If people behaved badly, he thought, one could always find a reason and try to help with it. "All that is necessary to prove this is to find out what the motives are for their superficial behavior, nasty, mean, or vicious though that behavior may be. Once these motives are understood, it is impossible to resent the behavior that follows." That unswerving dedication to decency—a foundation of the humanistic psychology movement—would make Harry's student famous, a hero in the counterculture 1960s, an influential psychologist even today, long after his death.

But in the early 1930s, Maslow was working at the Henry Vilas Zoo with Harry Harlow. Surprisingly, he liked it. He would come home, he remembered, wildly excited about some of the monkey experiments, so enthusiastic—like Harry, so suddenly tuned to the parallels to human behavior—that "my wife ferociously warned me against experimenting on her babies." Every day, Harry Harlow and Abraham Maslow were a little more impressed by monkeys and what they could do. There was something else—an almost unnerving awareness of a relationship. It wasn't just the monkey-to-monkey connection that impressed them. They were thinking about the relationship between the animals and themselves, the scientists and their subjects. It was slowly dawning on them that if one wanted an animal model in psychology, the smart, emotional, complicated monkeys might make a whole lot more sense than the maze-running rats.

This sense of kinship prompted Harry to ask hundreds of questions. What could monkeys actually do? What problems were they capable of figuring out? How close to human capabilities could they come? Maslow described his first assignment as running "a million boring delayed-reaction" experiments with monkeys, much like those that had so annoyed Tommy the baboon. The tests went like this: Maslow would show the animal some food and then, while the hungry monkey was watching, put the food under one of two cups. Then he would enforce a delay before the animal was allowed to search for the food. He would wait, one second, two, ten, thirty, sixty, whatever the design of the day proscribed. They wanted to know how good the animals' memories were. Suppose every monkey lifted the correct cup after a five-second delay. How about thirty seconds? Longer? And so on. And then they would analyze other details. Did age matter? Species? Sex?

Despite the nit-picking details, Maslow began to find the work addictive. The zoo-housed primates seemed to approach puzzles much as people would. If a monkey couldn't solve a problem immediately, it would start fiddling with the puzzle, continuing through trial and error, testing what would work. These experiments would continue until the animal found the error and fixed it. It appeared that the animals were thinking their way through the puzzles. They seemed remarkably goal-directed. They wanted—again like their human cousins—to win. They liked to beat the game. When they did, they would look up at Maslow, almost grinning. He would find himself grinning back: "I became fond of my individual monkeys in a way that was not possible with my rats," he wrote. And he liked Harry, who was only three years older, and more than open to friendship. They ate late dinners, talking over the work, trying to think where it might take them.

Maslow didn't like the University of Wisconsin. The ideas circulating in the psychology department seemed small to him, and the priorities equally as small. "The emphasis here is all on getting ahead," he complained. "Two articles are good; four are twice as good. It's all very mathematical apparently. There is a direct relationship between number of articles published and your 'goodness' as a psychologist." He thought that his professor, working in basements at the local zoo, showed the most promise. Harry Harlow, he wrote in his journal, had the makings of "a very brilliant man."

Of course, Harry also made Maslow work hard. Because Lewis Terman's standards were drilled into him, Harry insisted that Maslow make a rigorous study for his doctoral dissertation. When his graduate student protested that they'd already done innovative research—couldn't he just write that up?—Harry assured him in advance that the new work would be even more brilliant.

Maslow's dissertation was on relationships. Who has power, who doesn't? He spent hours at the zoo, watched thirty-five primates, from newborn to ancient, from spider monkey to baboon. He recorded every instance of what he considered dominant and submissive behavior. The resulting paper reads like a dictator's guidebook—and a testament to the way that hierarchy shapes social lives.

There was a clear sense of the system's winners and losers. Maslow pointed out that the power monkeys, the alpha males, pretty much had it their own way. They took food whenever they wanted it from other monkeys. They bullied subordinates. They initiated fights if they sensed a challenge. And they were always able to score sexually. The alpha male is pretty much guaranteed fatherhood and genetic success. Subordinate behavior was the near opposite. Male monkeys low in the hierarchy would cower when the power males strutted by. They were passive when attacked or muscled away from the females they had been courting. Their behavior was purely defensive; if challenged, they ran. This was hardly a position that promised a wonderful genetic future.

Maslow saw the ruler-and-serf pattern repeated across species. He concluded that the top-bottom structure defines most primate societies—monkeys, apes, and, although he did not say this, undoubtedly our own. In more recent years, primate researchers have marveled at the way a monkey pecking order can resemble a large corporation or a military hierarchy. If social order is controlled through dominance, Maslow said, even sexual relationships can assert power rather than affection. There's little doubt that the latter also represents a facet of behavior that unfortunately is observed in human societies.

Maslow showed that monkeys need beautifully tuned social skills to navigate these often risky social byways. He had been startled to see how savvy the monkeys were in dealing with each other. Primates in the wild could establish a relationship just by the way they looked at each other or gestured back and forth, he said. They could read body language. After studying each other, they might start a gentle grooming or they might just hurry the hell out of there. Such subtle exchanges served many of the animals well. It often allowed them, Maslow pointed out, to avoid bloody, destructive fighting.

If an alpha male could just intimidate a young male challenger into backing off, he didn't need to beat him away. That result might just delay the fight—but, for the moment, both monkeys remained physically intact. The zoo, of course, didn't have enough animals in its colony for researchers to observe the full social range of behaviors. But it was clear that a social primate definitely needed to understand a vast and potentially treacherous terrain of relationships. If you didn't like your place in the landscape, you needed to understand whether it was possible to move. For a young, ambitious monkey, the critical point in that decision would be challenging the hierarchy without, say, getting killed before achieving adulthood. If that didn't look possible, then, obviously, the key to a successful life was learning how to get along.

For most of us, Maslow concluded, getting along with each other was exactly the way to navigate through a long life—and, perhaps, even a happy one.

Harry was unabashedly proud of Maslow's work. In his view, the work on dominance was one of the best studies on power relations ever done. And Maslow had accomplished it, Harry pointed out, in a small zoo, working under a professor who had no research budget.

"Now that is creativity, when you can work with nothing and make a great scientific breakthrough," he wrote.

Even so, Harry was getting tired of working with nothing. He knew, Maslow knew, that monkeys were extraordinary animals. But almost no one else did. There was no network of federal primate research centers, although later Harry would help create that network. He no longer wanted a rat laboratory. He wanted a primate lab. To be taken seriously, he needed a place where he could gather his own animals, control his own experiments. Reporting out of the local zoo was not going to work indefinitely. It had been two years, by now, since Harry had arrived in Madison. The university still showed no interest in providing him research space of any kind. He was out of patience. If he was going to work in some outpost of science, a psychologist on monkey island, then he wanted a decent island.

The story of how Harry Harlow built the first primate lab at the University of Wisconsin still stands as a testament to determination—and deviousness.

In 1932, the university finally offered him an abandoned building. Harry described this gift horse as a "twenty-six-foot-square, two-story building on the wrong side of the Milwaukee Railroad tracks." The building was an old forest service property, built to test wood products such as crates and boxes. The interior was a maze of reinforced concrete posts. Some were slender spikes. The largest were sixteen feet tall and six feet by three feet at the base. The ground floor also held a tangle of disconnected pipes. The university would let Harry have the building as long as he didn't expect much in the way of remodeling money or help with the construction. He didn't care. It was a space, his space. "It looked awfully good to us," he said.

Harry persuaded a new graduate student, Paul Settlage, to help work on the building. Settlage was a friend of Maslow's. He was far less sure of where he wanted to go with psychology and more than willing to join in such a nontraditional approach to the field. Bearing sledgehammers, he and his professor marched up to the proposed laboratory. First, Harry wanted to clear out the forest of pipes and pillars. They managed to smash out a couple of small pillars, but the big ones barely cracked. The next day, Paul brought a cousin, Walter Grether, who was working on a degree in physics. The three of them chipped away a few more pillars. The next day, Paul and Walter brought pneumatic hammers.

Within a week, neither Harry nor his student helpers resembled anything like members of an academic department. They were gritty with concrete, dust, and sweat. By the time they had finished clearing out the factory—Harry estimated that they removed a good thousand feet of pipe—they looked like bodybuilders. Harry was struck by their newly bulging biceps and chunky shoulder muscles. "No matter how abstract our research became, it started out by being very concrete," he joked. By the end of the project, Grether had decided that psychology—or perhaps Harry Harlow—was a lot more interesting than physics. He changed his major.

"They just don't make graduate students like that these days," Harry once said, acknowledging that his early protégées not only studied at the primate lab, they built it. But when he and his team had finished remodeling the box factory, they immediately realized that it was too small. The university refused to approve or finance an expansion. Again, Harry was ready to get around that. There was a fair amount of land around the old building and one of his students had suggested that they might at least build some outdoor cages. Harry appealed to the university for materials to build these lightweight structures. This time, he won official approval.

It was all he needed. After all, Harry reasoned, the cages needed a concrete floor that could be hosed down. And you can't just pour cement straight onto the ground. So they put down six inches of cinders, four inches of crushed rock, and poured a good solid foundation. This was heavy work, obviously; but Harry had a few members of the football team in his class and he talked them into helping, "since they weren't doing very well in their studies." Watching the halfbacks and fullbacks heave sacks of cement around was "a beautiful sight," Harry said. They were in such good condition, he was sure they would have an undefeated season.

And as long as they were putting down such a good foundation, Harry thought they might as well put up good wall framing. Why not build those frames with studs that would allow doors and windows? He and his students covered the walls with insulating material. But then, well, Harry worried about the roof. After all, it snowed a lot in Wisconsin. So they put on a solid, sturdy roof. But then the well-insulated walls and the roof made the cages so dark that no one could see anything inside them. "There is no use in having an observation cage if you can't see in it," Harry said. So they cut out spaces for windows. As long as they were doing that, it seemed reasonable to open up the doors, too.

It occurred to Harry that the cages would get pretty hot in the sun. The only solution seemed to be to cover them with better insulation. So they covered the insulating panels with siding. "The drop siding cost considerable money. The only sensible answer seemed to be to give it a coat of paint." And then it turned out that the tarpaper on the roof kept peeling up. So they re-covered the roof with asbestos shingles. The shingles happened to have a twenty-year warranty. The new structure was so close to the lab building that it made sense to add corridors connecting it to the laboratory. The lab crew could do that with leftover concrete and wood scraps. "When we finished, we were horrified to see how much these Outdoor Observation Cages looked like real buildings."

The university was horrified, too. Harry received what he called a "very sharp note" from the comptroller saying that it was absolutely illegal to build wooden buildings that didn't meet state specifications. Harry was unfazed. He had just loaned the university president a monkey (as a pet for the president's son), and the chief administrator himself had picked it up. Harry and his students were sitting on the roof, driving in nails at the time. "After we put on our shirts, we had a very friendly chat with the president. As far as we could tell, he wouldn't have cared if all his staff built laboratories."

So Harry wrote to the comptroller explaining that this was just a project that had gotten out of hand. And, by the way, it needed some electrical wiring, steam heat, overhead lights, and good ventilating fans to really be up to code. While the university electricians were connecting the fans, Harry persuaded them to wire in floor plugs. The following year, his new extension appeared on the official campus map.

The laboratory now resembled nothing so much as a ramshackle house with a deeply overgrown yard. Harry dug into his own pockets again and paid for some shrubbery to give the laboratory—and the monkeys—some privacy. The researchers planted ivy and grapes along the fence, honeysuckle bushes, wisteria, lilacs, forsythia, pine trees, and poplars. "In the summer, we couldn't see out and others couldn't see in."

It was a cheerful, informal place. Everyone had to do odd jobs to make it work, including the professor. If students were short of money, Harry let them unfold cots and sleep there. One summer, Paul and Walter, almost broke, moved into the laboratory and made their meals out of bread and fish caught from the campus lake. They occasionally shared tidbits with the monkeys. One black spider monkey named Gandhi loved the fish so much that they eventually let him join them at the backyard table for lunch. "By the end of the summer, Gandhi had table manners that would have been a credit to a Harvard man," Harry said.

He was building up his small colony of monkeys: spider monkeys, like Gandhi, who were agile South American tree dwellers; a little group of capuchins, another South American rainforest species, once famed as organ grinder monkeys. There were Asian monkeys, too, especially sturdy rhesus macaques from India, who would eventually become the primary lab dwellers. The monkeys were built differently, colored differently, and regarded each other with deep suspicion. They seemed equally suspicious of their captors—but not hostile. Harry once accidentally locked himself into a monkey cage and only escaped when three passing sailors, home on leave, heard him yelling and pried the door off its hinges. The monkeys had apparently considered Harry an extremely odd cage mate because they gave him as much room as possible during the entire episode.

Occasionally—too frequently—the animals escaped themselves. They could free themselves from cages far more adeptly, Harry noted, than researchers. In his early days, he said, they often had at least one monkey in a tree. On one occasion, half a dozen macaques terrorized a small neighborhood near the campus for more than a week. They raided restaurant kitchens and threw acorns out of trees at passersby until they were finally trapped after trooping through a window to explore a second-floor apartment.

At this point, Harry had wonderful stories to tell—during the week-long monkey escape, he had received a letter from Canada advising him to get the animals drunk—and some genuinely compelling studies of monkey intelligence. Mostly, he had small-scale tests at the zoo. Now, he had a place to show off what he—and, more important, his monkeys—could do. He had big plans for the systematic, controlled studies that would convince the behaviorist, rat-model-trained psychologists who surrounded him. Now that his laboratory was completed—shabby, patched together, but there—he had every intention of tackling some of the mysteries of the thinking brain. Of course, he still wasn't sure exactly how to do that.

FOUR

The Curiosity Box

It is my belief that if we face our problems honestly and without regard to, or fear of, difficulty, the theoretical psychology of the future will catch up with, and eventually even surpass, common sense.

Harry F. Harlow, 1953

AT THE LITTLE LAB IN Madison, Harry had a group of three capuchins, those limber, bright-eyed animals once known as organ grinder's monkeys. He named them Capuchin, Cinnamon, and Red. God, they were smart.

Even in so small a cluster, the three males formed one of Maslow's dominance hierarchies. Capuchin was the boss monkey, Cinnamon second, and Red third. Capuchin was not a nice monkey. He was a bossy, greedy little food-hoarder. He took the best treats for himself. He took the others' scraps, too. He would share with Cinnamon, but not with the lowly Red. Red, perpetually hungry, was driven to plotting for his meals. He would creep cautiously up, when Capuchin was busy, and sneak back his stolen dinner.

One summer morning, Harry and an anthropology student, Leland Cooper, were standing somewhat idly by the capuchins' outdoor cage, when Red came by on a crumb-foraging expedition. Up stepped Cinnamon, the big fat tattletale, screeching out threats and yelling for Capuchin. To all appearances, Red then lost his temper. He grabbed a stick from the cage floor and gave Cinnamon an angry poke. And when the alpha monkey, Capuchin, muscled his way over, Red whacked him, too, even though "he had never been known to use a stick for striking at any time prior to this."

Once having realized the weapon's potential, though, Red didn't forget it. Cooper later reported another incident in which Red and a fellow capuchin were sharing cage space with five burly rhesus macaques. Macaques are bigger, tougher, and meaner than capuchins. They slid into bully mode, forming a circle around the two smaller monkeys. According to Cooper, Red picked up a stick again and started whistling it through the air around him. But the macaques were too quick for him to reach and they leaped resentfully out of range. Still, they stayed out of range and left the little capuchins alone after that.

One of the most interesting aspects of the story of Red, the stick-wielding capuchin, is how long it took Harry Harlow to tell it. The observations were made in 1936. The report was published in 1961. He explained that "publication delay resulted from the authors' reluctance to report this unusual observation until they had achieved established reputations." Animal intelligence was an oxymoron when Harry Harlow was building his primate laboratory. This was the day, after all, of the conditioned response and the simple and reflexive brain. For an animal to reason that a stick was a useful weapon would suggest thought and calculation. A scientist who reported that kind of cognitive ability in monkeys in the 1930s was likely to be branded a sloppy observer or a wishful thinker. Or both.

It was a rare moment of caution for Harry. Perhaps even caution is too strong a word, for some calculation was also involved. He knew that monkeys were smarter than the profession would admit. The trick was figuring out how to prove that. Plenty of scientists before him had based their arguments on similar anecdotes and failed to sway the crowd. Nineteenth-century proposals for animal intelligence had been dismissed as sentimental or as based on anecdotes rather than evidence. Some very good early twentieth-century psychologists had done studies showing strong evidence of problem-solving abilities in chimpanzees without reversing the general prejudice against intelligence in other species. They included the respected American psychologist, Robert Yerkes, and Wolfgang Kohler, a German gestalt psychologist who had done a famous series of experiments in the early 1900s when he put chimpanzees in a cage with bananas dangling overhead. To reach the bananas, the apes had to figure out that they could stack boxes, which were tumbled in the cage, and climb them. Kohler had argued that this was a genuine "Aha" moment, that chimpanzees were capable of insight. Kohler's work is heralded today. At the time, though, he struggled to make his point. The leaders in behaviorist psychology accused him of superimposing human behavior on another species. As Harry's student, Abe Maslow complained, successful psychologists wouldn't even listen to the argument: "It is now fashionable to despise Gestalt psychology," he wrote in his journal. "Accordingly, they all despise it."

It would be easy for Harry to become more of an outcast than he already was. To make his case, he needed more than good monkey stories, clever anecdotes. There had to be a way to devise a believable intelligence test for monkeys, something systemic, something objective. He was working cautiously in that direction when two events drove him more directly into the fray: He went to New York. And he lost his temper.

In 1939, Harry received a one-year fellowship in anthropology at Columbia University. The Harlows moved to Manhattan for the academic year. The call from New York was perfectly timed; Harry had that restless, itchy, something-around-the-corner feeling about his work. Clara was expecting their first child. She decided to take the months in New York as an opportunity to enjoy being a mother and to think about what might come next. She wrote to her mother, with typical determination: "I have a feeling that a job will get me again but not until we have a firm hold on family plans. I do not agree with women who take six weeks off to have a baby. I want first to know my own child thoroughly so I will know what parts of his life to leave to others and what to keep management of myself." Robert Mears Harlow was born on November 16, 1939, and both Harry and Clara were mesmerized. The baby, according to Clara, was just "irresistible," and Harry was spending extra hours at home to admire him.

At least he was until Kurt Goldstein came to lecture.

Goldstein was one of the great European neurologists of the day—intense, brilliant, passionate. His research blended concern for mental health with hardheaded clinical study. A native German, he had worked long and desperate hours trying to help soldiers with head injuries after World War I. Goldstein's experience with brain damage had led him to try to understand how the brain was organized so that he could learn how to repair it. He had patiently tested injured soldiers, seeking to determine which head injury produced which specific failure of memory or motor skills. What kind of damage twisted numbers around? What made words vanish?

Goldstein had found that the brain-damaged soldiers were more rigid and inflexible in their responses. They could do what he called "concrete" learning—rote memorization, simple recitation of stored facts. But ask them to reason through a problem, such as change the order of numbers, the pattern of the shapes, and the soldiers struggled. They seemed almost paralyzed by the shift in perspective. His patients had lost their "abstract attitude," in Goldstein's terminology. They were unable to adjust their answers. Their thought processes seemed to have stiffened and become "concrete," he said.

Early in 1940, when little Robert Harlow was just a few months old, Columbia scheduled a series of lectures by Goldstein. The old neurologist promptly began talking about his famous division of concrete and abstract intelligence. He went beyond brain injuries, though. He used the same dividing line to separate humans from the other primates. Goldstein had never been fully able to accept Darwin's evolutionary ideas, the notion that the brains of humans and other animals might have common origins. At Columbia he declared flatly that monkeys sat on a lower rung of intellect. They could accomplish rote learning, he said, but nothing complex, and never abstract reasoning. Monkeys were born to be no more than the brain-damaged soldiers in their abilities. The other primates were concrete learners. Only humans could achieve analytical intelligence.

Harry sat through those lectures in a state of increasing disbelief. Goldstein was an inspiring teacher, Harry said, but he was absolutely and completely wrong about other primates. Harry had now spent eight years watching monkeys. He knew that they could reason their way through a problem, rethink a challenge. Wasn't that exactly what Red had done, when defending himself with sticks? And there were countless others, from Jiggs and his puzzle work to Tommy's pleasure in getting the right answers. Back in Wisconsin, Harry had another monkey that he considered a natural engineer. In one simple test, that capuchin had matched Kohler's chimpanzees when he balanced sticks and boxes against the side of a building to reach food that the scientists had cleverly dangled from the roof. Harry found himself indignant on behalf of his animals. If his teacher really believed that monkeys possessed only concrete thought processes, Harry wrote to a friend, then Goldstein was "a cement wit" himself.

When Harry came home at night, he walked the baby up and down in the small Riverside Drive apartment, talking, pacing. Baby Harlow's nighttime lullaby was a litany of the history of psychology. The cocktail hours with Clara were filled with discussion of wrong-headed science; indeed, "work was the background music of our lives," one of Harry's children would later recall. Harry picked his way through Goldstein's arguments. He was out of patience finally and completely with the rat-psychology view of the world, with the simple brain and simple behaviors, with ignorance and prejudice toward other species. It seemed to him that by dismissing the abilities of other species, in the end, psychologists were dismissing the abilities of their own.

He knew how deep the counter-arguments ran. It wasn't just that Romanes and Kohler and other distinguished scientists had failed to persuade. It wasn't just that Watsonian behaviorism and Pavlovian conditioning were dominant. Scientists had been insisting for centuries that animals were basically brainless. The other species could be conditioned, they could be made to respond; but think, feel, analyze, grieve—never. Back in the 1700s, French philosopher René Descartes had likened animals to machines; animals could never think as humans do, he said. They were soulless creatures, beast machines. That perception held even when Charles Darwin made his evolutionary arguments. Darwin undeniably suggested that humans and other species must share common brain structures and therefore common abilities. It was too much for Goldstein, who responded by dismissing evolution outright. But even those who believed in Darwin often could not quite accept that animals possessed the kind of complex brains that had long been reserved for humans.

The idea of intelligent animals had a particularly rough time in the United States. One of the most famous books on the subject, *Animal Intelligence,* published in 1898, basically concluded that animals weren't intelligent at all. The author, New York psychologist Edward Thorndike, tended to side with Ivan Pavlov. Animals could be trained—or conditioned—to look intelligent. But that, Thorndike said, was misleading. His most famous test involved putting cats into boxes and testing their ability to escape. The boxes were small enough to make the cats feel just a little squeezed, a little antsy to get out. Thorndike provided them with an escape mechanism. The boxes had panels that could be opened when the cats pressed a button or pulled on a string. To reward the cats for opening the panel, Thorndike placed a food treat just outside the box. To strengthen the intensity of that reward, he kept the cats hungry. His experiment involved measuring the length of time it took a cat to break free.

After some time in the box, the cats would push, bump, and eventually trip the string or step on the button. The next time in captivity, the captives would move more directly to the button. The more often the cat went into the box, the faster it got out. After a few trial runs, some cats were pulling the string almost as soon as the box was closed.

Some people would call those cats smart. Thorndike concluded almost the opposite. The feline behavior showed no evidence of thought, he said, merely "the accidental success of the animal's natural impulses." Thorndike went on to develop "laws" of animal behavior. His Law of Effect came directly out of the cat-in-the-box work. It said this: If a movement is followed by the experience of satisfaction or the removal of annoyance, that movement will be "connected" with the solution. In other words, if the cat pulls the string and the box opens, eventually it will connect string-pulling with box-opening. His second law, The Law of Exercise, said that the more this happens, the stronger the "connection" between action and result. In other words, the cat will become a string-pulling automaton. Thorndike first called this somewhat robotic turn of events "stamping in" behavior. He later came to prefer the word "reinforcing," a term still used in psychology and animal training today. He considered his laws comparable to the laws of motion and energy in physics, another step toward making living creatures as predictable as clockwork.

The mechanical animal—incapable of love or reason—obviously fit well with the teachings of early behaviorists such as John B. Watson. But it got an even bigger boost from Harvard-trained researcher Burrhus Frederick Skinner, perhaps the most famous psychologist of Harry's generation. Known as B. F. Skinner to most of the world and as Fred to his friends, Skinner was adamant in his belief that animals do not have feelings. He was appalled once when, watching a squirrel gobble a nut, a friend remarked that the animal "liked" the acorn. Of course it didn't, Skinner replied. Animals don't like things; liking is an emotion, and squirrels don't have those. Skinner described himself as a neobehaviorist, a builder of a more sophisticated version of the earlier science.

In pursuit of that ideal, Skinner created a device that became known far and wide as "the Skinner box," an updated version of Thorndike's apparatus. The square box was soundproofed and equipped with a bar or lever. If a rat pushed the lever, a food pellet tumbled out. If a pigeon pecked the bar, it, too, received food. The rodents and birds pushed and they pecked and they ate, just as Skinner had predicted, in the most convincing way. During World War II, Skinner was able to use his box to train pigeons to peck at a target. He tried, unsuccessfully, to persuade the U.S. Army that the birds could be put into the nosecones of missiles and used to guide weapons. Of course, if the food-delivery mechanism jammed—which it sometimes did—the pigeons rapidly lost interest in pressing the bar or pecking the target. After all, what was the point of thumping on cue if no food came out? From a scientific point of view, Skinner appreciated this reluctance. It was, as he pointed out, a classic Pavlovian extinction curve. But Pavlov's beautiful calculation of vanishing behavior made army officials doubt the reliability of pigeons as bomb-delivery systems.

Harry Harlow was not a fan of the Skinner box. It tended to inspire him to sarcasm. "There is no other learning technique that ever did so much for the pigeon," was his summary. It was a relief to learn, he said, that if his brain dwindled to pigeon-like dimensions, he could still be conditioned. Meanwhile, a pretty good pun occurred to him: "It is nice to know that it takes little brain to learn or think, and as I grow progressively older, I am enormously reinforced by this discovery."

It wasn't that Harry denied the veracity of such experiments. He'd done classical conditioning studies. He knew they worked. He just didn't think that such responses were *everything:* "Our emotional, personal and intellectual characteristics are not the mere algebraic summation of a near infinity of stimulus-response bonds." He had his allies, especially among the young skeptics like himself. Harry was sharing his ideas with a new band of scientists that included the Canadian psychologist Donald O. Hebb, renowned today for his farsighted theories about how experience influences the brain. Back in the 1940s, Harlow and Hebb were, at best, promising outsiders. In exasperation, Hebb once declared that it would be better to be wrong about how the brain worked—to stand behind some real ideas—than to be as vague and inconclusive as psychology was at the moment.

Measurable behavior—not the black box of the brain—was still in fashion. Thorndike and Skinner were mainstream. The most important psychology theorist of the time, Clark Hull of Yale University, was building an entire system of behavioral predictions based on the concept that stimulus and response were defining characteristics. People listened to Hull. He was a soft-voiced, articulate, and dedicated scientist, liked and respected by many of his peers. According to one analysis, 70 percent of all studies dealing with learning and motivation during the 1940s cited one or more of Hull's books and papers.

Hull's theory was an almost geometric structure, crystalline in its sharply defined architecture. Often called "drive reduction theory," it was based on the idea that behavior is created by drives or needs that we seek to satisfy or reduce. Hull's overall concept had seventeen corollaries, seventeen postulates, and assorted theorems, proofs, and formulas. A classic Hullian equation might include stimulus (s) and drive (D) and response (R) and habit strength (sHr) and number of reinforcements (N) and hours of food deprivation (h). All this might add up to a formula such as

$$sHr = h \times N \times R.$$

In other words, the strength of a habit equals hours of food deprivation multiplied by the number of reinforcements (amount of food) and by response to that situation. In other words, a very hungry rat, which receives food when it presses a bar, will develop a very strong bar-pressing habit. The hungrier the rat is, the stronger the habit.

The message was a familiar one. The bar-pressing rat was a conditioned animal, responding to a hunger drive rather than exhibiting intelligent thought. It was Thorndike and Watson and Skinner all over again. But Hull's theory seemed to his colleagues to take psychology to the next level. It integrated the experiments, pulled together the results, in a systematic way. It also followed the classical notions of science: putting forth theories that could then be tested.

Leading experimental psychologists, such as Kenneth Spence of the University of Iowa, turned their labs and their students over to testing those elaborate calculations. "Hull's theory was truly scientific and so I became a Hullian," explained William Verplanck, an emeritus professor of psychology at the University of Tennessee, looking back toward his student years in the mid-twentieth century.

Did Harry Harlow go for this? Not in the least. Spence and Hull used to exchange exasperated letters about Professor Harlow, who didn't believe in the theory and appeared, in their opinions, to be needlessly outspoken on the point. Harry didn't step back. He made sarcastic remarks about people who thought the "Hull truth" was the whole truth. "Harry was no theorist," Verplanck said. "He was simply a hardheaded empiricist. He just followed his nose and published what he found out." Over the years, Verplanck, who died in 2003, came to agree with the Harry Harlow perspective on theorizing, and perhaps even goes beyond it: "I think that theory is the curse of psychology—if we get rid of the theory, we might know something."

Actually, Harry was less troubled by the theories in general than this one in particular. He thought Hull's idea depended on studies that made animals look simpler and stupider than they really were. He had come to think that popular experiments, even the Skinner box, achieved little in understanding how an animal's brain handles complex situations. Did psychologists really believe that they could define behavior with the discovery of the "empty organism," simply managed by simple training techniques? Surely there was more to the brain—and to us—than that?

It wasn't that Harry opposed testing for animals' abilities. It wasn't even that he was opposed to building devices that could be used to look for a particular behavior. He was—like Skinner or Watson or Thorndike—a dedicated experimentalist. He believed in the power of evidence gathered in the laboratory. But he also believed that too many psychologists were setting artificial limits on their subjects. How much intelligence did it really take to press a bar or to push a button? One of the standard mazes of the time was a T-shape. A rat could hurry down a long straight arm and then turn either right or left. How much of a challenge was that?, Harry asked. How much did one really learn from watching a rat run forward? Was there value in proving that rats can move fast? A device was needed that would really challenge animals to think. He didn't want a one-time challenge such as Kohler's box-and-banana problem. He wanted a systematic way to push monkeys to achieve, beyond the techniques of the scientific community. He wanted to take the way Goldstein tested his human patients and apply those standards to monkey intelligence.

When he and Clara and Robert returned to Madison, they had two different projects planned, one personal and one professional. The Harlows were going to build a new house. Clara was going to oversee that project. And Harry was going to hasten over to his shabby laboratory and design a thought-provoking device, one that met all the scientific specifications. It's worth noting that the first paper out of Harry's lab, describing a "Test Apparatus for Monkeys," was recommended for publication by that device-loving psychologist, B. F. Skinner himself.

Eventually, the test apparatus was formally named the Wisconsin General Test Apparatus, or WGTA. In primate research, it became a genuinely famous design. Copies of the WGTA can still be found, mechanized and modernized, at primate centers around the world. When one of Harry's later graduate students, Allan Schrier, took a job at Brown University, he was still proud of having worked with the original. Schrier purchased vanity plates for his 1966 Volkswagen bug that said 66WGTA. When he bought a new car two years later, he updated his license plates to 68WGTA. The older plates went to Harry with a note: "Rhode Island recognizes a good apparatus when it sees one."

Here's how the first WGTA worked: There was a cube-shaped cage, two feet in all dimensions, with a solid oak floor, three-quarters of an inch thick (hardwood floors were cheap in those days). A sliding panel that could be raised and lowered with a rope-and-pulley system was fitted to one side of the cage. A monkey would sit in the cage, waiting, presumably a little curious about what was going to happen. When the panel rolled up, the animal could see a table edged with brass rods. The researchers could slide trays onto the rods and ease them right up to the monkey. The trays were packed with test objects and treats. The monkey could reach through the bars of the cage to grab, discard, and puzzle over test objects; and, of course, pick up treats. When a monkey had worked through one set of challenges, the scientist could replace the tray. At the opposite end of the table from the monkey was a small observation post. There, tucked behind a one-way screen, the scientist could watch the monkey without being seen himself. The WGTA was a good design and that was one of the few things that B. F. Skinner and Harry Harlow agreed upon.

What made the WGTA look brilliant was something else. Harry still didn't have enough monkeys in his lab. There were no domestic breeding colonies. Monkeys were hard to find, expensive, and often, after being trapped and shipped in less-than-nurturing conditions, half-dead when they arrived. He considered his few dozen healthy animals solid primate gold. He hoarded them. Out of simple preservation, he had to throw out the standard rules of animal testing, which were based on rodent work. Rat research worked on the principle of unlimited supply. When psychologists were testing conditioned responses, they often wanted inexperienced rats for each study. If an animal was already conditioned in one experiment, it was hard to separate the effect for the next study. So rats were rarely recycled. Harry once described the standard psychology experiment as a Blitzkrieg involving forty-eight rats: "The controls are perfect, the results are important, and the rats are dead."

Harry Harlow might have taken the same approach if he'd had a similar river of monkeys flowing through his lab. But he had only a small pool, one he couldn't afford to drain. He was a psychologist who had a finite number of test subjects and an infinite number of WGTA tests that he wanted to conduct. One forty-eight-monkey "do and die" study would have left him with a lab full of empty cages. He never even considered it. So instead of conducting a Blitzkrieg, he rotated his monkeys, four at a time, through those countless studies. Over and over again, first one problem set, then something harder, then something harder yet. Unlike the rats—and even the cats in previous studies, the monkeys couldn't avoid building on previous experience. The result was that the WGTA didn't just make monkeys look smart, it made them look like small geniuses. To the surprise of everyone, including the Wisconsin psychologists running the tests, the monkeys began making educated decisions, fast and savvy. "Had we run many monkeys on just a few problems, as was the custom of the rodentologists with rats, we never would have realized that animals could learn how to learn," Harry said.

That gave him an opportunity, which he took, to argue against the rapid killing of research animals. Harry reminded his fellow psychologists that they could benefit by keeping their animals alive: Even if you did believe that all human behavior could be worked out in rats, he said, the "do or die" design of standard experiments was just flawed. It couldn't explain people very well because people don't generally do one task at one time and then fall dead. The practice of psychology might open up a new understanding of very short-lived rodents, he suggested, but it didn't do much for anyone, human or other animal, who lived long enough to gain a little experience.

The scarcity and cost of monkeys also forced Harry to rely on the cheapest primate possible, the abundant Indian rhesus macaque. That, too, turned out to be an extraordinary piece of good luck. Rhesus macaques are not the most beautiful of monkeys. They lack the elegant ballet dancer build of a squirrel monkey or the endearing fluffiness of a titi monkey. Rather, rhesus macaques look competent and tough, steelworker monkeys on their way to the factory. Wiry golden gray fur frames a squared face dominated by a long snout and a pair of close-set eyes, coffee-dark and alert. Primarily forest dwellers, they can live almost anywhere. They have the rare ability to take advantage of wherever they land; they skitter through city streets, colonize old temples, raid farms, whatever it takes.

Rhesus macaques thus are resilient, adaptable, and, perhaps even more important to researchers, accessible. Scattered across such a range of habitats, they are easy to find and therefore an ideal research monkey from the collecting point of view. Even today, rhesus macaques are the most used monkey in research. Modern medicine owes them much. The Rh factor in blood (positive, negative) was worked out in these monkeys. The Rh, in fact, comes directly from the term *Rhesus.* During the desperate 1950s race to find a vaccine for polio, scientists tested rhesus macaques by the boatload and by the cargo plane–full. So many monkeys were scooped out of India—well over a million—that the country banned their export in the early 1960s, fearing soon none would be left.

Harry hadn't at first realized how smart the macaques were. Their ability to adapt is, like ours, based partly on a quick, calculating mind. Today, primatologists have shown that rhesus macaques can do simple math equations and play shoot-the-target arcade games with astonishing accuracy—far beyond that which even Harry Harlow expected. So Harry was lucky twice over. He was lucky to have had so few monkeys and lucky that the monkeys he could get were mostly rhesus macaques. And later, when he went on to study love and connection, he would find himself lucky again. Because rhesus macaques are also, like us, among the most passionately connected species on the planet. Once again, they would help make Harry Harlow a psychologist with the right animal at the right time.

The first tests with the WGTA were straightforward. Most intelligence tests that people take are written. A few tests, for young children and brain-damaged adults, are administered with pictures or objects—a picture of a block, an actual block. Goldstein had helped develop such tests. The monkey tests were all based on objects set out on those interchangeable trays. For example, a monkey is shown a board or tray that holds two objects. Let's say there's a fat blue cube and a long green rectangle. The board is flat but there are regularly spaced hollows in it—as with the board of the African stone game, Mancala.

Researchers call those hollows "food wells." The experimenter can put the cube and rectangle on top of the well to hide food or treats within. If the monkey picks up the "right" object, say, the cube, it finds raisins or peanuts underneath. Under the rectangle, no such luck. Essentially, finding the treat provides the clue to the right answer. The cube and rectangle are moved around the board here and there, to the corners, in the middle, to make completely meaningless patterns, and the monkey has to try again to pick up the right object. The cube. The cube. The cube again. Trial after trial until he always grabs, yes, the cube.

And then a new trial, and now it's the rectangle. Or there's a new pattern, a blue ball and a spiky purple cross. This time it's the cross. And then a different trial. And a different trial.

So how did the few monkeys, picking up blocks over and over, become important? If this was only trial and error, then each time the scientists changed the pattern, the animal should go through a comparable fumbling process toward the answer. If it took it thirty trials to get the cube every time, then it should take the same thirty or so trials when a researcher changed to the rectangle or the cross or the ball. But that wasn't what happened. Instead, the monkeys got faster and faster, and more and more accurate. After a monkey had spent some time in Harry's lab—and that could mean hundreds of tests in a week or so—it could figure out the pattern within one or two tries. If the treat was under the blue cube, that was the answer. If during the next test the blue cube produced nothing, but there was good stuff under the red triangle, the monkey would rethink. At first, the animal might take a few trials to recognize the switch, maybe six or so. But the more experienced monkeys could shift from cube to triangle after one mistake, almost as fast as the scientists could change the trays. "Eventually, the monkey showed perfect insight when faced with this kind of situation—it solved the problem in one trial. If it chose the correct object on the first trial, it rarely made an error in subsequent trials." And if it did pick the wrong object, it immediately changed over. As Harry put it: "This is not a vague something-

or-other but a definite and measurable concept on the part of the learner as to what is being learned. It is the point at which a child might say, 'Oh, I'm supposed to *add* these numbers and I know how to do that.'"

In fact, the animals were starting to make this look too easy. Harry and his students decided to make the challenge harder. Now they offered the monkeys a choice among three objects. Harry called this an "oddity trial," and it was the kind of test that Goldstein believed could prove analytical ability. The tray now contained three food wells, two covered by matching objects and the third covered by a misfit object. In these tests, only the odd object sits on top of food. So let's say that in the first test there are two blocks and a funnel. The animal must choose the funnel. Then the scientists switch it—two funnels and a block. Now, the correct object is the block. There's a subtle and important distinction here that the animal must understand: It is not the shape of the object that is important, but its relation to the other two objects. In the first test, the funnel shape means odd one out. In the second, the funnel shape means matched set. The problem of shapes and relationships is complicated because it demands an ability to analyze relationships. And, again, this turned out to be no problem. After a short period of puzzled experimenting, Harry's monkeys again came through like stars on the oddity trials.

They were so good that Harry started wondering about another comparison: monkeys versus humans. How would his bright-eyed macaques compare to other beginners on these kinds of tests? Say, for instance, young children? He became so intrigued by that idea that he recruited a child psychologist to work with him, an upcoming new faculty member named Margaret Kuenne (pronounced KEE-nee). She arranged for a group of children to try solving the same kinds of block-and-funnel problems. The children were rewarded with beads and small toys rather than peanuts and raisins. The researchers deliberately decided to select smart children. Peggy Kuenne put together a group of seventeen children, aged two to five, who had relatively high IQs, between 109 and 151. The children had no previous experience with oddity testing. Both children and monkeys were then asked to solve puzzles. Both groups fumbled through the first few problems, but they gradually recognized the pattern and whizzed through the later tests with near perfection. "Sometimes the monkeys were better, especially early on. Once both were trained, the children were almost always faster at picking up the pattern but the process was basically the same," Harry wrote. He was definitely making his case that monkeys shared analytical abilities with humans. He wondered, though, whether he had even yet pushed the animals as far as they could go.

If the monkeys could understand relationships between related objects, could they understand even more subtle symbolism? Harry and his students laid out a complicated series of relationships on the WGTA trays. Now the color of the tray mattered. On one test, the tray contained three objects: a red U-shaped block, a green U-shaped block, and a red cross-shaped block. If the tray was orange, the monkeys had to choose the green block, recognizing that it was the odd one out in color. But if the tray was cream-colored, the monkeys had to choose the cross, the odd one in shape. This was not something the monkeys did first time out, obviously; but with practice, wrote Harry, "after the monkeys had formed these two learning sets, the color cue of the tray enabled them to make the proper choice, trial after trial, without error."

This was exactly the kind of problem that Goldstein had found was so difficult for people suffering brain damage. And now Harry started thinking again about the absurdity of setting limits on the brain. His WGTA monkeys were learning, becoming smarter with education. The academically challenging life of these monkeys seemed to be extending their natural abilities. So what if Harry took on Goldstein even more directly? If monkeys with a whole and healthy brain could gain analytical skills, what about brain-damaged monkeys? Could the injured brain also have the potential to improve? Harry happened to have some animals that might be just right for trying out this newly heretical idea.

Several monkeys in his laboratory were left over from a study of brain anatomy. These monkeys had lost one of their two brain hemispheres from an early effort to explore the right versus the left hemisphere. Due to his monkey-hoarding principles, Harry had kept these animals well fed and healthy. He decided to put the half-brained monkeys into the WGTA tests. Again, because he had only a few of them, these "hemidecorticate" animals learned the lessons of the trays and the objects and the shapes and the colors over and over again. And, against all the predictions, they became measurably educated monkeys. Although the half-brained monkeys were slower than their peers, they still gained speed and accuracy as they rotated through the test trays.

Harry's team compared the "educated" half-brained monkeys to uneducated but intact macaques. For a brief shining period, until the new monkeys were also well educated, the brain-damaged animals looked smarter. In the early comparisons, they were far faster at the tasks. "It would appear that in this situation, half a brain is better than one, if the trained half-brain is compared to the untrained whole one," Harry wrote. He went on to speculate that there could be the possibility of therapy in these results. "More seriously, these data may indicate why educated people show less apparent deterioration with advancing age than uneducated individuals." As an educated man, Harry appreciated the idea, but it was another point that he thought more important. His results emphasized to him that no one is simply born brilliant: "The brain is essential to thought, but the untutored brain is not enough, no matter how good it may be."

It was an absolutely wonderful idea—that we all had promise, we all had the potential to be coaxed out by the right teacher, or home, or even laboratory. If our brains weren't good enough, they just hadn't yet received what they needed. For all that it was a B. F. Skinner–approved apparatus, Harry's WGTA ended up as pure counterculture in its results. It didn't show conditioning at all. It showed learning. Harry's monkeys didn't look anything like bar-pressing rats or target-pressing pigeons. Naturally, the results came under attack. And one of the criticisms was troubling. All the animals in these experiments were being given food rewards. Didn't that make them like Skinner's pigeons? The monkeys might look sophisticated, but perhaps they were just being conditioned in a different way? If so, there was nothing new here besides Professor Harlow's fancy ideas.

Now, in a psychology experiment, a food reward is partly a way for a scientist to bridge the communication gap. When testing a human subject, we can give stars or grades to indicate correct answers, we can cheer and congratulate. But how do you tell a monkey that he's done it correctly? The Wisconsin scientists, in the time-honored tradition of animal research, used food treats. That certainly allowed critics to argue that, despite Harry's evidence, despite the monkeys' increasingly adept test taking, the animals were really responding to a food stimulus. There was still room for the interpretation that he was just turning out primates who were well-trained but not especially bright.

If Harry Harlow really wanted to prove that he was demonstrating complex brains at work, he would have to answer that criticism. And so he started thinking about motivation itself. There are certainly the well-trumpeted survival drives—hunger, thirst, shelter, safety. But people, Harry believed, are motivated by other, equally powerful drives. And it's those other drives that lift us beyond basic existence. We are propelled by emotions such as love or anger, or by that illusive pursuit of happiness. We are driven, too, by a sense of wonder, of exploration, by courage and curiosity. And it was the last of those, the itchy, pushy, irresistible force of curiosity that Harry now began to consider.

He had a story that he particularly liked about the naturally curious monkey. One evening, Harry had been working late at the lab. When he was ready to leave, he remembered that building maintenance had been complaining about lights left on in the empty building. Harry carefully flicked the switches to the off position, went out to the parking lot, got into his car, swung it around to leave—and saw that the lights had flashed back on. He stopped the car, went back in the building, and turned the lights off again. But this time he didn't go back outside. He just stood there, quietly in the dark, waiting. Abruptly, the lights blazed on again. That was when he caught, out of the corner of his eye, a tail sliding beneath the bars of the nearest cage. It was Grandma, one of his spider monkeys, and she was clearly entertaining herself by flicking on the lights with her long, prehensile tail.

But how had she figured it out? And why? She didn't receive food or water or strokes for pushing a switch with her tail. Harry was almost as delighted by Granny's light-flicking maneuvers as he was by the detailed experiments performed with the macaques. If that wasn't successful curiosity, what was? She'd been curious, she'd figured it out, and her reward was a new skill—and the ability to light up the lab—nothing more or less. It seemed to him that figuring out a puzzle, solving a question, could be its own reward, even for monkeys. The old orangutan, Jiggs, brought this to mind with his passionate desire to master the square-peg dilemma. It had never been food rewards that interested Jiggs. He'd wanted to beat the game. Harry, Peggy Kuenne, who was still working with him on the tests, and his increasingly fascinated crew of students, were finding that food rewards—the bedrock of behaviorist conditioning—often seemed surprisingly irrelevant. At least to primates.

This idea was so heretical that Harry decided once again that they would need an apparatus to pursue it. What they came up with was a mechanical puzzle: a wooden block on which a hasp was restrained by a hook, which was restrained by a pin. The hasp, the hook, and the pin had to be opened in precise sequence to open the puzzle itself. They put together two groups of four monkeys. In one group, each animal was given the puzzle but no rewards. In the second group, each animal was given the puzzle and then rewarded with raisins for a correct move. Conditioned response theory predicted that only the raisin-fed monkeys could be trained to solve the puzzle. But their results went exactly in the opposite direction. The monkeys who were offered no food did stunningly, obviously better.

Put simply, mixing food and puzzles distracted the animals. They were happy enough to get the raisins and peanuts, certainly. But if the monkeys were hungry, they had a hard time focusing on the puzzle. Some of the monkeys grabbed the raisins, stuffed their faces, and then sleepily lost all interest in the challenge. Those who weren't hungry were happy to get the treat, but they would wedge the fruit into their cheek pouches, saving it, and continue puzzling over the hasps and the pins. The monkeys who solved the puzzle most efficiently were those who had no food distractions. They merely sat down and went to work. You could turn these results over and inside out. There was no way to conclude that the researchers had conditioned a puzzle-solving response into the rhesus macaques.

If you really think about the learning-for-food idea, Harry said, it makes no sense anyway. We're primates ourselves and if primates learn only to satisfy hunger, then few people in the well-fed United States would have an incentive to learn anything. People and monkeys alike learn because they are curious or interested—and that can be more potent a force, on occasion, than a wish for a second slice of pie. Harry could hardly wait to tell his colleagues about his monkeys, their cheeks puffed out with food treats, still puzzling over a tricky set of locks.

"He had this enormous and contagious enthusiasm for research data being collected in the lab," says Robert Butler, a postdoctoral researcher at the time and later a professor of psychology at the University of Chicago. When he first came to Harry's lab, Butler had begun to think that he'd chosen the wrong profession; his early psychology classes seemed completely flat. "I had been wondering why I went for my Ph.D. in psychology. It was so boring. Psychology was so defensive about being a respectable science by doing experimental work. But Harry made you see it differently."

Harry put Butler to work doing delayed response trials with the WGTA. While he was waiting the proper delayed time period—thirty seconds or so—Butler became curious himself. He started to wonder what the monkeys were doing while they waited for the sliding panel to open. So he added a mirror, angled so that he could watch the animals in their cages. He hadn't considered that the monkeys could also see him. And he certainly hadn't considered that it might matter if they did. Suddenly, though, his test results started falling off. The monkeys were fumbling through their challenges. Finally, Butler realized that the monkeys had lost interest in the trays because they were more interested in him. Instead of sorting blocks, they were watching that strange face in the mirror. For all the years that scientists had been finding monkeys fascinating, no one had thought that monkeys might find scientists equally interesting. These monkeys were abandoning the food rewards on the tray just to study a reflection in an angled mirror. Butler started thinking, once again, about the curiosity experiments.

So he built the first scientific testing box for Harry's laboratory. Butler's solid-sided box had two moveable windows, one red and one blue. It was designed so that a monkey inside could hear noises outside the box but not see what made them. Would they wonder about what was out there? If the monkey pushed the right window—picked the color correctly—the window would slide open for thirty seconds and he could peer out at the world around him. That glimpse was the only reward for picking the correct color. Did he open the window? Is the earth round?

In one experiment, a monkey persistently raised that colored panel from sunrise until the last grad student left the lab at night. In another test, Butler alternated a plate of delicious fruit outside the window with the chug-and-toot of an electric train. Food was, sure, food; but the train became an obsession. The monkeys studied the fruit with undeniable greed. But the train—so completely strange—riveted them. They couldn't figure out what it was. They needed just one more look. The windows flew up and down like winking eyelids.

When Butler first proposed the box, Harry was doubtful. But, as Harry frequently said, he was often wrong. That was exactly what Harry told Robert Butler: He thought the box wouldn't work, but try it anyway. And when Butler ran the tests and the results were electric with curiosity and nothing else, Harry was "ecstatic," Butler said. The idea was so smart and the results so good that some of Butler's friends warned him to publish in a hurry. After all, major professors had been known to take credit for their students' ideas. Instead, Harry plastered Butler's name all over the findings. He named the device "the Butler box." "I didn't name it," Butler relates. "Harlow named it because he wanted some opposition to the Skinner box." In Harry's mind, the Butler box was the perfect counter to rats pressing bars. Butler's invention demonstrated, without a doubt, that animals were curious. They had thinking minds of their own and they used them, whether researchers dangled food bait in front of them or not.

Harry liked to point out that this was a beautiful example of science catching up with everyday common sense. It was his favorite kind of psychology—the kind that made sense in the real world as well as in the laboratory. As he said, "An informal survey of neo-behaviorists who are also fathers (or mothers) reveals that all have observed the intensity of the curiosity motive in their own child. None of them seriously believes the behavior derives from a second-order drive. After describing their children's behavior, often with a surprising enthusiasm and frequently with the support of photographic records, they trudge off to their laboratories to study, under conditions of solitary confinement, to study the intellectual processes of rodents."

Harry liked the Butler box so much that he kept it, even when he was no longer himself doing intelligence studies. It would turn out to be a smart decision. When he became interested in mother love, Harry put baby monkeys inside the box and their cloth-soft surrogate mother outside. He knew that electric trains and strange scientists fascinated monkeys. None of that compared, though, to the way the baby monkeys would doggedly raise the panel to see their mother's face. The little animals would open the window and open it and open it, over and over and over, until one by one the graduate students watching them dropped off to sleep. In one experiment, a baby monkey continued to seek those flickering glimpses of his mother for nineteen hours straight. It might have been longer, actually, but yet another student fell asleep as he watched the window flick up and down. The baby monkeys were so fixed upon seeing their mothers that Harry took to calling the Butler box a "love machine."

You couldn't watch the small monkey faces, their eyes anxiously searching for their mothers, without beginning to see love as a tangible force, a physical cord pulling tight between mother and child. You might even come to believe that love is so powerful that it can influence anything, including the brain. You might, if you were a scientist watching those monkeys, start thinking that the tireless blink of that window, the serious little face peering through it, had something to tell you.

FIVE

The Nature of Love

Growing up is very gruesome / by singletons or else by twosomes / And after love has long miscarried / The twosomes find that they are married.

Harry F. Harlow,
"The Gruesome Twosomes," undated

ALREADY A FEW OTHER REBELLIOUS scientists were arguing that love and intelligence could be connected, literally, from dot to dot. These were not animal researchers but doctors and psychologists working directly with children in orphanages and foundling homes. They suggested that social intelligence and cognitive intelligence might be linked. The end point was that children raised without affection might lose more than their ability to relate to others. Isolation and loneliness might dull the brain in other ways—and that dimming down might even show up on the Stanford-Binet Intelligence Scale.

To those opposing the idea—and Lewis Terman was definitely among them—the concept was ridiculous. Sentimental. And unnerving. If the healthy development of the brain depended on being loved, wouldn't that suggest that affection and nurturing were akin to breathing, basic to life itself? If that were so, wouldn't we be impossibly vulnerable to loneliness or isolation or the vagaries of parents and home?

Terman didn't see the brain as anything like such an unstable structure. The human brain gleamed in his mind like finished marble, sculpted by genes, polished by superior biology. If you had inherited good genes—the kind possessed by Clara Mears Harlow—you possessed "splendid hereditary equipment" and were born to be smart. If not, you were destined to be average, or worse. Children who inherited the good genes looked good on IQ tests. Children who inherited less-splendid genetic material didn't. Terman's vision didn't allow for a smart child who faltered in a hostile environment. It allowed little room for environmental influences at all, including the difference between a warm and loving home and an unfriendly one.

By such standards, a slow-maturing baby could be judged harshly. If an infant seemed behind the curve, some pediatricians would advise that the child be institutionalized. The parents could then try for a better one. Arnold Gesell, one of the best-known pediatricians of the 1930s, had a reputation for recommending that approach. Gesell, a Yale University psychologist, remains famous in baby-doctor circles even today. He was a pioneer in working out developmental timetables—all those charts that dedicated parents follow in anticipation of when their babies should start smiling or rolling over. And Gesell used to say that the inborn development tendencies were so strong that parenting styles didn't matter that much. A child was going to turn out as his genes dictated, so that he "benefits liberally from what is good in our practice, and suffers less than he logically should from our unenlightenment."

Gesell opposed the adoption of very young babies. He thought prospective parents should wait to see whether a child suffered from inborn brain deficiencies. It didn't occur to him that being institutionalized might foster deficiencies. But when he evaluated those older orphans, he quite often found them less sharp than the parents had hoped. Another Yale psychiatrist, Milton Senn, complained that Gesell's habit of diagnosing mental defects in institutionalized children kept them from being adopted at all. Senn recommended early adoption as a preventative measure *against* mental retardation. Gesell had a ready answer for that: He accused Senn of not understanding child development.

And yet there were these annoying studies that kept turning up here and there. In New York City, one outspoken psychiatrist was making the most inexplicable findings about affection and IQ. The researcher in question was William Goldfarb, who had been studying children in the city's Jewish foundling homes, trying to assess their social and intellectual development. Goldfarb had been one of the first researchers to worry that social isolation could permanently affect children's ability to connect with other people. He also tracked the IQ scores of children raised in foundling homes and compared them with the scores of children raised in foster homes. His findings directly challenged the notion of superior genetic lines.

The mothers who had left their children in the foundling and foster homes had to fill out an education and background survey. "It is a matter of some importance," Goldfarb insisted, "that the mothers of the institution children are significantly superior to the mothers of family [foster] children in occupational background." With this statement, he was deliberately taking on the Terman point of view, the belief that superior parents produced superior children. Just in case anyone had missed his drift, Goldfarb hammered it again. The mothers of many children placed in institutions came from the higher social classes. Their children were the result of unplanned and unwanted pregnancies. By contrast, many of the foster children came from a less impressive family background. They were placed in foster homes due to neglect, the death of a parent, desertion.

If you followed the laws of inherited intelligence, you would expect that the children given to foundling homes would inevitably be smarter than the fostered children. After all, Goldfarb said, "One might even infer that the mothers of the institution children are also superior in intelligence." What he found and reported, though, was the opposite. Over all, the fostered children averaged 96 on the IQ scale. The foundlings averaged 72, falling into the dreaded feeble-minded category. The foundlings were less determined, less interested, less willing to explore. What could have happened to these children of supposedly bright and capable mothers, then? Goldfarb thought they were probably diminished by the sterile, unnourished nature of the homes. The places were stripped down, after all, and understaffed. The children were raised in an atmosphere of clean rooms, carefully ordered play, many domestic chores, and very basic instruction. It was hard to imagine that a child would thrive intellectually in such a world.

Goldfarb, though, also worried about another kind of diminishing effect—less obvious but, he suspected, no less real. He reported that many of the foundlings were so apathetic that they appeared as shadows of children. They were silent and withdrawn. Some could hardly be tested because it was so difficult to awaken them into focused participation. One problem was that no one was interested in them, he said. The caretakers seemed indifferent. But was that surprising? Goldfarb asked. Is an adult ever interested in a child who doesn't stir his heart? An odd kind of chicken or egg issue underlies that query. Does affection for another person create interest in him or does interest lead to affection?

When it came to the foundlings, Goldfarb had an idea that interest and affection twined together, tight as a rope, almost inseparably. All of us, even as babies, are a bundle of feelings and desires, he said. Our positive emotions grow best in an interactive sense, fostered by how we react to others and how they respond to us. A baby, a child, even an adult, needs at least one person interested and responsive. We grow best in soil cultivated by someone who thinks we matter. A baby, in particular, needs such encouragement and will do his best to please in return. Infants imitate adults and coo back to them and smile back, and through those ordinary exchanges they have their best chance at developing into an engaged and confident child. Without such affectionate interaction, Goldfarb thought, those positive responses would fail to flourish. The exterior child would look healthy; the interior would be stunted. Lacking a strong caretaker relationship, a child surrounded by other children in an orphanage could still grow up in a kind of curious developmental isolation.

Goldfarb believed that foundling children had a "never sated craving for affection." Because no one cared about them, they buffered themselves by not caring either. They withdrew from others and they withdrew from tasks and challenges, including those that you might consider intelligent life skills such as reading, and math, and those analytical challenges built into the Stanford-Binet IQ test. So what did their test scores reveal? Mental deficiency or intellectual despondency?

That question was taken up at the University of Iowa's Child Welfare Research Station. In the late 1930s, the Iowa psychologists were focusing—ahead of their time—on the interaction of genes and environment. Perhaps Goldfarb's closest counterpoint there was a famously gentle-mannered psychologist named Harold Skeels. His research had started with a focus on language development. Skeels had also been following children raised in foundling homes and testing them on the Stanford-Binet scale, tracking their language skills, as they grew older. What intrigued—and worried—him was the same thing that had troubled Goldfarb. Skeels wasn't seeing the normal rising curve of language skills that he had expected. The longer children stayed in the homes, the more their verbal IQ scores dragged downward.

No matter how he turned those results around, the one constant was that the foundlings felt desperately unwanted. Skeels also began to wonder whether a lack of loving attention could impair intellectual functioning. He tried a simple test. He took a group of preschool children from a warehouse-style orphanage and sent them to a friendly nursery school for several hours a day. Skeels then compared them to toddlers who stayed at the foundling home entirely during the next year. The children who did not attend nursery school outside the foundling home suffered the usual IQ drop. As he reported, those who went to preschool didn't suddenly leap upward in verbal

IQ. But they didn't tumble, either. Their scores held steady. Compared to the foundlings, still sliding down the scale, the preschool group thus looked a lot smarter.

The nursery school, Skeels reasoned, provided only casual affection. What he needed—if he was right and affection mattered—was to provide something like mothering. Because the children's mothers had long vanished, he needed a good loving substitute. This led to Skeels's most unusual—some might say risky—experiment. He took thirteen children, all under the age of two and a half, from an orphanage and put them in a home for older "feebleminded" girls, those who fell below the razor wire Stanford-Binet line of 80. Skeels carefully selected girls who were clearly functional and warm in nature. Each child was "adopted" by one of the older girls, a few by an attendant, who took over mothering functions. And mother they did: They cuddled and kissed, played with and comforted the children in their care. Over nineteen months, the average IQ of the mothered toddlers rose from 64 to 92 on the Stanford-Binet, in other words, from feebleminded to measurable intelligence. There was indeed something, still mysterious, about isolation that seemed to make the brain falter.

But what was it? Later in his career, Harry Harlow would take on the effects of social isolation as directly as anyone in psychology. For the moment, the issue merely hummed at the edges of his awareness. The power of love and loneliness was an interesting academic question, even a troubling one. He wasn't ready to take it on; he was still obsessed with questions of working intelligence. But, as it turned out, the effects of isolation were about to gain his attention in a personal sense. He might not be ready to study loneliness—although that, too, would come—but he was heading for a sharp lesson in the shape of life without love.

We can create isolation in institutions. We can, it seems, also create it while still surrounded by family. At this moment, Harry's professional life seemed to him to be the best thing in his life—his intelligence studies were gaining recognition. Terman was finally impressed, even predicting that his former student would rise to the top of the American Psychological Association. It would be many years before Harry would admit that he might have been wrong about what was best in his life. After he retired, he would talk about how hard it was to get love right. We could be told, we could be educated, but we still had to fumble our way through the lessons of the heart. The challenge would lead him to the rare admission that it might take a greater power than even science. "People have to learn quite a bit by themselves," Harry said simply. "Christ passed on to people quite a few tantalizing tidbits about the importance of love and left the rest of us to learn, little by little, as God sees fit."

Harry's oldest son, Robert, remembers those years when his parents' marriage began to break as almost a crazy quilt, a patchwork of good times and bad. The Harlows had built their house on Lake Mendota, a rectangular building made beautiful by windows filled with the glimmer of light on water. On weekend mornings, Harry and Clara would have coffee together. Bobby, as Robert was called then, would sit near his parents in a little wooden chair that had belonged to his mother and had her name painted on the back. Robert still calls it the Clara chair. He would hold as still as he could, a small, fair-haired five-year-old in the Clara chair, listening to his parents. "If I was good, they'd let me come over and each give me a spoonful of coffee. It must have been full of cream because it was very mild." He was a quiet little boy, undemanding, good enough to get coffee regularly, and his mother used to call him her perfect child.

Clara was still a lover of friends and company. The Harlows hosted bridge and dinner parties. Bobby would sit on the stairs, "listening to the sound of bridge cards getting shuffled." He holds onto other memories besides the dry whisper of cards. He recalls swimming in the lake during the summer months; making his mother an ashtray out of tar from the driveway, which melted into a black sticky puddle; of his dad's aversion to house and lawn work. There was one week when the whole family was felled by illness. Paul Settlage wrote to his friend Abe Maslow, back in New York, that "recently, Harry has been having a tough time of it. Bobby had the mumps, and then Clara caught it, and then Bobby became sick twice more, and during this time, Harry discharged the domestic duties while being more or less in the toils of flu himself. The other day he could hardly stand up. The man always has had a certain dogged persistence."

On December 10, 1942, Harry and Clara's second son, Richard Frederick, was born. Bobby's memories become less happy after that point. It was as if that change, just the one more child, pushed Harry past what he could handle. He had no spare energy. His research was becoming an obsession, the laboratory seemed to be his home of the moment. Settlage saw this as a visible change, a shift into total absorption. "Harry," he told Maslow, "is working harder than ever." There was a new driven intensity. And with new demands at home, Harry seemed to feel stretched too far—like a fine wire, thinning as it pulled. It was barely six months after Rick was born, according to Clara, that Harry began to withdraw from the family. He "increasingly immersed himself in his work and became silent and uncommunicative" at home. He was up early and gone, home late, off to the lab every day, weekends included.

Finally, Clara began insisting that, at least, Harry should take Bobby with him on weekends. He should have some time with his son, she argued. Even lab time was better than no time. Now grown, a father himself, Robert Israel still remembers those visits to the box factory lab: "He'd take me over there and I'd watch him do experiments, slide the door up and down, arrange the puzzles. I loved being there. I could wander around and anytime I wanted I could feed the monkeys. There was a container of dried fruit and another of peanuts and I'd get a handful. He taught me how to hold the food so I wouldn't get scratched." They didn't really do the father-son thing that Clara had envisioned. At the lab, even with his small son in tow, Harry's mind was only on the research. "He didn't talk to me. When he was at the lab, he was focused on his work. But I was comfortable because Dad was there. I was happy in his world. My brother, who's three years younger, I don't think he has a single memory of Dad from his childhood."

In a furious, and later regretted moment, Harry told Clara that he didn't know whether he loved her anymore. He wasn't sure that she loved him, either. He wasn't sure she ever had. He threw the question at her, suddenly angry over the fourteen years of their marriage. He seemed like a stranger to her, she said. The house on the lake was becoming a place that anyone would want to avoid. Harry and Clara were barely speaking to each other; quiet little Bobby was tiptoeing around the house and Rick, now a toddler, was studying his father as if he couldn't quite remember who he was. Around his friends and colleagues, Harry suddenly became silent about his marriage. His conversations were only about work. His letters were bright and talked only of professional issues and achievements. After a series of such communications, a puzzled Lewis Terman wrote and asked him why he never mentioned Clara in his letters any more. Had she left him? Harry replied with another letter full of psychology news.

Clara filed for divorce on August 14, 1946. Her pleading with the district court is a litany of bewilderment and grief: Harry was coming home later and later, skipping dinner with the children, even when she begged him to give them some time. He was showing up late for social engagements, embarrassing her. He was impatient with her and impatient with Bobby and Rick. He had "developed a practice of ignoring and rebuffing" inquiries made of him by Clara or by either of the two small boys. She was living in silence and hostility; she was worrying constantly; she couldn't watch her children being pushed away like this by their father. It wasn't that she wanted out so much as that she couldn't stay.

Harry did not defend himself. It was a rare moment for him—he refused to fight back. Clara won custody of both boys—not unusual in the 1940s or even today—and a less usual uncontested division of property. The lake house was appraised at $20,000 and put on the market, the proceeds to be split equally. After the mortgage was paid off, and closing costs deducted, they each had $7,473.46, to the penny. Take whatever else you want, he told her; and in her anger, her worry and grief, she wanted all of it. She took the furniture and lamps and cushions and rugs and artwork, the stove, refrigerator, washing machine, vacuum cleaner, dishes, glassware, silverware, everything, according to the divorce settlement, except Harry's clothes and personal effects, such as hairbrushes and handkerchiefs. They owned $1,000 worth of AT&T stock and an $800 war bond; he sold all of it and gave her half the cash. He agreed to take over a $5,000 life insurance policy and to continue paying the premiums on it. He agreed to pay all legal costs. He agreed to pay $150 a month child support for three years and then $100 a month until the boys came of age. He agreed that the children could visit him each year. He agreed to pay the costs of the visit. He agreed to accommodate her schedule in the visit. He agreed to work with her on making the journey safe from parent to parent. The divorce went through in a flat three weeks; it was final on September 6, 1946. Clara left Madison almost immediately and moved to Rhode Island with the boys to stay with her brother, Leon.

"Dear Abe," wrote Settlage to Maslow. "Did you know that the Harlows were divorced recently? It was quite a shock—totally unexpected by us but apparently suspected by others for some time. I had the impression that the Harlows were getting along more congenially as time went on. Quite a psychologist, am I not?"

Harry was alone, for the moment, with psychology as mistress and wife and family. No inconvenient children, no messy marriage cluttering up his life. It didn't take him long to realize that he hated it. Up close and personal, the field of psychology was a less than rewarding companion. There was nothing in it, especially at the moment, to help a man come to terms with a failed marriage and a silent home. Harry had a small apartment again and plenty of time, in these bright, open, empty days, to pursue his research and to realize just how chilly his profession had become. Perhaps nothing exemplified that better, at the moment, than his own department at the University of Wisconsin.

It had been a long time since Harry's department was crammed into the basement of Bascom Hall. The psychologists now occupied a premium place on the shores of Lake Mendota, thus commanding a glittering view and some of the leakiest, dampest facilities on campus. They had rat labs now and those were in a basement that "flooded with every rainstorm, so you had to wade to your equipment. Not that deep but it wasn't great to be standing in water with all that electrical work," recalls psychology professor Richard Keesey.

Perhaps the chronic damp affected the mood at 600 N. Park, the department's slightly unfortunate address. If a sender had scrawled the direction at all, the address on the envelope tended to look like GOON Park. It seemed to the occupants that the mistake happened frequently. "The mailman always knew right where to deliver it though," Keesey says, raising an eyebrow with deliberate irony. Goon Park became the department's unofficial nickname, partly because it seemed to reflect the uneasy politics of the place. There were faculty members who didn't speak to each other, faculty members who accused their colleagues of academic theft, faculty members who spent their days making sure that everyone else knew their places in the hierarchy, who made sure that only those on the approved list could even have coffee in the department lounge.

The famed psychologist Carl Rogers is still remembered, decades later, as one of the unhappiest members of the old Wisconsin department. Rogers created the idea of client-centered therapy. His point was straightforward: Psychologists don't always know more than their clients; therefore, therapists should actually listen to their clients. Widely accepted now, it was initially a strange and unwelcome idea. Many psychologists resisted Roger's call for open-minded counseling. They were the ones who had trained as experts in human behavior, after all. At Wisconsin, a department of dedicated experts, Rogers sinned further by aligning himself with the humanist psychology movement. By the 1960s, Rogers and Harry's former graduate student, Abraham Maslow, would both be leaders in that movement, arguing that in psychology the emphasis should always be on human potential rather than negative emotions and neuroses.

In retrospect, it seems obvious that Rogers was a poor fit for the Wisconsin psychology department at mid-twentieth century. He was talking about compassion and decency at a time when the department was still following the Hullian model of mathematical behavior. Those who weren't math-minded were often treated as substandard. Rogers complained that the department conspired to make people such as himself—and most of the students—live under a sense of perpetual threat. Instead of attending faculty meetings, Rogers started leaving a tape-recorder that was set to play his comments in his absence. In a 1964 memo to the department, shortly before he ended his seven-year stay, Rogers assured his fellow faculty members that he could no longer stand the place. He accused the Wisconsin psychology professors of being obsessed with methodology and finding fault with others, "both of which constitute further insurance that no significantly original ideas will develop."

Even before then, others were beginning to worry that Wisconsin's approach to behavior had dried out. People began referring to the department as "the dustbowl of empiricism," and they were only half-joking. To graduate, students ran a gauntlet of extreme mathematical calculations. University of Oregon psychologist Michael Posner, a former assistant professor in Madison, recalls: "Each student had a methodology examination that was required for the degree. I was assigned to grade these exams. One year when I saw the exam in advance, it included a very complex Graeco-Latin square experimental design that I had never heard of or seen talked about in the literature. I remember searching desperately for a reference and finding a single obscure paper by the department chair. So I was able to grade the papers but the students, of course, did not have the advantage of seeing the questions in advance . . . I guess they were just supposed to know these things. Needless to say, exams were rather tense situations."

Harry found the department's mathematical obsession not just wrongheaded but boring. Of course, Harry literally had nightmares about numbers. He told friends that he suffered from a recurring dream that Stanford had called him to tell him that he'd never really gotten that Ph.D. because he'd failed the department's statistics exam. He rejoiced in his experiments that used only four monkeys at a time. Simple statistics, the kind anyone can understand, he liked to argue, "are almost as powerful as common sense." Common sense happened to be one of Harry Harlow's standards for good science.

"At Wisconsin, when you developed a dissertation, it was supposed to be very systematic," says Michigan psychologist Bob Zimmermann who earned his Ph.D. in Harlow's lab. "So I decided, okay, I'm going to do a systematic study of brightness discrimination in the monkeys. I came up with this beautiful plan. Black and white squares, dark gray squares, light gray squares, very systematic, very statistical, very Wisconsin. And Harry looked at it, and said, 'That's the fastest way to obscurity, obviously.'"

Goon Park wasn't all statistics obsession and unfriendly behavior, of course. The psychologists were a social community. They hosted dinners for colleagues, picnics, cocktail parties. There was enough friction that hosts had to exercise some caution in invitations. One of the Wisconsin psychologists lived in certainty that his colleagues were stealing his best work. "Academic bandits," he would shout, his accusations echoing in the halls. Faculty wives of the time remember planning their invitation lists to avoid meetings of the different factions. Harry partied—he could drink with the best of them—but he was beginning to see himself as an outsider yet again. He didn't leave tape-recorded messages, but he became less visible at faculty meetings. And less visible in general. "He wasn't unfriendly," says one former colleague. "Just not friendly." He attended parties, but not all of them and not with memorable enthusiasm. One of Posner's memories of Harry Harlow is of the eminent psychologist snapping at him at a party, asking for another drink, and obviously not enjoying himself. "I wasn't as unhappy with him for asking," Posner says, "as I was with myself for getting the drink."

Harry was increasingly on his own, at his now-empty home and in his notably tense department. You wonder whether emotional isolation can change the child, rearrange the brain a little? It can also change the adult scientist. That period of intense loneliness and of disconnection would make Harry tougher and sharper. He would never again be as visibly sweet as Clara Mears had found him. He would turn more to sarcasm as a defense. He had fewer drinking buddies and he would compensate, over time, by drinking more alone. And sometimes he would feel alone enough to be downright hostile about it. When colleagues in the psychology department complained that Harry's battered laboratory was too much of a private empire—and they did complain—he responded diplomatically by putting a sign on the building that read: "Department of Psychology Primate Laboratory." But he was less diplomatic when asked about the sign. "He said he did it 'to make the bastards happy,'" says long-time administrative assistant, editor, and friend, Helen LeRoy.

If he was going to have a support system, it was clear, more than clear, that Harry Harlow would have to rebuild his private life. He was learning a lesson that he would later prove experimentally in haunting detail: We aren't meant to be alone. Isolation is only a punishment. Social species—and we are undeniably that—thrive only in a garden bed of relationships and connections. Not all of us need large gardens, not all of us need traditional families. Most of us—and this comes right out of attachment theory—need at least one good bedrock relationship.

What Harry missed most was marriage as partnership. By now, he knew that he needed someone—as Clara had been at first—who could be a partner inside psychology as well as outside of it. When he'd married Clara, that shared interest had illuminated their relationship. He was a man "who woke up thinking about his work," says a long-time friend. He needed to be with someone who accepted that, even appreciated it. He didn't have to look far. His fellow researcher, Peggy Kuenne, was right there. And she, too, was looking for a smart partner.

Peggy was a pretty woman, shining dark hair and clear blue eyes in an elegant face. She had pale skin, fine, high cheekbones, and a generous mouth that she liked to paint with bright red lipstick. "I recall a lot of men who tried to date Margaret," her brother Robert Kuenne says. They were attracted by her good looks and quick mind. She was rarely attracted in return, though. "She worshipped one thing and that was intelligence. Her turndowns of men were brutal; she never could tolerate a fool." He remembers how impressed his sister was with Harry Harlow. She told her brother that when Harry gave a lecture, intelligence gleamed right through the words.

The oldest of three children, Peggy was raised in a working-class family in St. Louis. Their father was a compositor for the local paper; their mother was a milliner. Both parents expected their children to achieve far greater things. The Kuennes pushed the children and they watched over their schoolwork. It was easy for the parents to watch. The family lived in a tiny bungalow built by the father. Later, their middle child, Robert Kuenne, would wonder whether that enforced intimacy had turned them all into people who craved distance from others: "There was very little privacy and we were all very private people. Independent and inner driven."

The Kuenne children became a trio of high achievers. Robert went to Harvard, earned a Ph.D. in economics, and was recruited to Princeton, where he spent his career as an economics professor. Dorothy, the youngest, became an atomic physicist at Washington University in St. Louis. And Peggy blitzed through her master's degree at Washington University and then went to the University of Iowa to study Hull's theories of conditioning in children under Kenneth Spence. "She was very interested in rigorous data, in showing that psychology was scientific," her brother recalled. Peggy graduated in 1944, became an assistant professor at the University of Minnesota. Two years later, she took a job at the University of Wisconsin and joined Harry Harlow's research team. They were natural collaborators, and after Harry's first marriage fell apart, their relationship shifted almost effortlessly into something more intimate.

A year and a half after his divorce, Harry Harlow married Margaret Kuenne. The ceremony took place on February 3, 1948, in the tiny town of Anamosa, Iowa. The out-of-state wedding was a strategic move. "I guess you would say they eloped," Helen LeRoy says. Iowa was the place for quick weddings then—it didn't require the blood tests and waiting period that Wisconsin demanded. A couple could just slip across the border almost invisibly; Harry didn't invite his Iowa family to witness the marriage pact. There were strategic reasons for that secretive ceremony and they all had to do with the University of Wisconsin. The school still enforced its rigid nepotism policy, the same one that had forced Clara out of her graduate program and into a department-store job. By eloping to Iowa, Harry and Margaret Harlow hoped to slip undetected under the university's radar.

They returned and continued to work together. They published together. At work, they treated each other with cool professional courtesy. They didn't go so far as to pretend they weren't living together. The newlyweds rented a small apartment near campus. Neighbors still remember being invited by Harry to have a drink and listen to his research ideas. It was inevitable that the news of their marriage would eventually filter out; when it did, the university's reaction was exactly what they expected and feared. It didn't matter that Peggy was already a fully trained psychologist and had been hired on her own merits. The administration insisted that one of them must leave the psychology department. Neither Harry nor Peggy considered that the person to quit would be him. "They both wanted Harry to be famous," Robert Kuenne says. Once again, a wife of Harry Harlow stepped down at the University of Wisconsin.

Peggy had professional advantages, though, that Clara had lacked. There was that Ph.D. in psychology and that reputation as a very smart scientist. Peggy thought of herself as a psychologist still. So did Harry. He thought she was too good a scientist to waste. This time, at least, he was prepared to end run his employer. He gave Peggy an office in the primate laboratory and she became the lab's unofficial editor. She spent hours polishing Harry's papers and those of his students. When Harry's old professor, Calvin Stone, retired as editor of the *Journal of Comparative and Physiological Psychology (JCPP),* he named Harry as his successor. Harry took over the journal in 1951, two years after his second marriage, and promptly recruited his second wife to help him edit it. Harry always said that Peggy was the more ruthless editor. He told his students that it sometimes took him weeks to persuade her to approve what even he had written. Once, he hid a paper that she had rejected; when he showed it to her again a month later, he said that he had rewritten it. She liked it the second time around. This proved, Harry said, that the occasional memory failure could be a good thing.

Harry's students remember the contrast the Harlows presented when they walked down the hall together: Harry, slight and a little scruffy, Peggy, straight and slim and neat, her head topping his. Mostly they worked in their separate offices. When he'd annoyed her, though, her voice carried sharply through the hallways. "She'd sort of screech out his name when he'd done some editing work she didn't like," says Bob Zimmermann, trying to imitate her call. Zimmermann's voice rises into a sharp falsetto. "H-a-a-a-r-r-y! . . . I wish you could have heard it."

The students and staffers at the lab found Peggy very different from Clara, who used to pack picnic lunches for Harry's grad students. Friendly informality was not her style. It hadn't really been her family's style. Harry called her Peggy. Her parents and siblings called her Margaret (and when she was a child, "little Margaret," after her mother). Her style at the lab was cool and formal, sometimes to the point of brusqueness. Some of the young scientists, used to a warmer welcome at the lab, were both hurt and put off by her manner. "An ice bitch," recalled a former student flatly.

Harry was called, on occasion, to be the mediator between the students, the staff, and Peggy. Sometimes the disagreements even spilled out of her office, where she kept a little desk lamp that allowed her to work in a small, warm pool of light. Peggy always hated the cold brightness of fluorescent light. When she had to share her office with students, it turned into a battle with those who wanted overhead lights. California psychologist Bill Mason, a postdoctoral researcher in the lab, remembers Harry trying to referee one particularly angry dispute and walking away with some relief. "I think I came through that pretty well," he told Bill, ignoring the mutterings following him down the hall.

He and Peggy had two children. Pamela Ann Harlow was born on September 20, 1950. Peggy was unreserved in her delight in her daughter's arrival. "Years of training, both experimental and clinical, have not deprived Peggy of the privilege of thinking that Pamela is 'real cute.' This is no doubt correct, though she looks to me like a baby girl," Harry wrote to Terman. On the same letter, Peggy scribbled a far more sentimental note: "As Harry implies, I am completely won over by our month-old daughter. She's a lively baby and has learned some things a little faster than her parents would like, but we wouldn't want too 'good' a baby." Peggy was equally excited about their second child, Jonathan, born three years later.

In that brief baby note, you can see a warmer, sweeter side to Peggy Harlow than generally showed in the lab. There she worked hard to maintain professional dignity. At home, she could put that away. She could laugh and play with her children. By most accounts, Peggy Kuenne Harlow was selective in her relationships. She loved a very few people and she saved her energy for the people who mattered to her most. She wrote regularly to her sister, Dorothy. But it was Harry who took over most of the correspondence with Peggy's mother. The Harlow files at Wisconsin still contain a thick packet of letters to the elder Margaret Kuenne and they all begin, "Dear Mother." And they all close, "love, Harry." Even now, Peggy's brother, Robert, recalls how kind Harry was to their mother. And what a wonderful mother his sister Peggy was to her children. "She really loved those children," he says.

Peggy wasn't a natural homemaker. It wasn't, after all, her first interest. She didn't know how to cook when she got married. But she clipped recipes and studied them and learned. She planned healthy meals (although when the children were sick, she brought them ice-cream sodas to eat in bed). She didn't try to keep the house immaculate. Visiting students recall dusty furniture and piles of papers everywhere. The papers might be expected because she and Harry both brought work home constantly. There were usually school projects spread out in one corner of the home or the other. "There was never enough time to fight that uphill battle for tidiness," Pamela says. Today you would think of the Harlow house as the typically cluttered home of working parents with children, a family where you choose your domestic priorities. The house might not be polished to 1950s homemaker standards, but Peggy always made time for her children.

When he could spare the time, Harry appreciated the challenges of parenting. He just thought they were a lot funnier than Peggy did. He loved to tell friends about his efforts to teach Jonathan how to put on a seat belt in the car. Child psychologist Dorothy Eichorn, of the University of California at Berkeley, was one of Harry's closest professional friends. She still remembers him chortling over his parenting skills. "He had this story he would tell on himself about when seat belts first were installed in cars. And he was trying to train Jonathan to use them and they'd gotten in the car and he said to Jonathan, 'Now, what's the first thing you do when you get in the car?' and Jonathan said, 'Shut the door.' And he thought that was just wonderful." Eichorn thought it was Harry's ability to laugh at himself that made it easy for him to charm a child. "He had a wonderful sense of humor and he knew how to capture children's interest. He taught my son to pitch pennies—not that I was so happy about that. But he enjoyed himself and I think he enjoyed children."

Harry and Peggy purchased a comfortable, 1920s house near the zoo and he would walk his daughter down to the park on clear nights to star-spot. Neither of them were experts on the constellations but they always picked out the Big Dipper. Harry taught Pamela how to whistle and to sharpen pencils with a knife. She no longer whistles, but she still makes pencil points with a knife. Some Sunday mornings, he would prepare oranges for the children in grapefruit style, cut into neat spoonable sections. He taught his children to play bridge, Jonathan remembers, and sometimes played tennis with them. He attended school picnics and plays during lulls in the research agenda. "But that was about it," Jonathan Harlow said. "He really didn't do much for fun. My father really spent morning, noon and night at the lab. He would walk home for dinner and then go back to the lab and work. My mother was a child psychologist and I think she really wanted children and my father had had children before with Clara and I don't think he was that interested in us."

Of course, this was the 1950s. How many fathers were that engaged with their families and housework then? It was mothers who made the costumes and coddled the sick children and spent their extra hours at home. But there is a real point in Jonathan's comment. During his first marriage, and then when he was on his own, Harry had discovered that there are other, less traditional ways of making a family. And it was his other family—the intellectual one—that often occupied him more, and sometimes seemed nearer and dearer to his heart.

He built his other family at the laboratory. It contained an extended network of graduate students, and post-docs, and dedicated staff, and he loved being there, in that family, as much as any place in the world. Personal relationships could fail, as he had learned, and disappear. But down at the primate laboratory, there was always a person willing to listen, and work to be done, and a sanctuary from the math-minded discussions of his colleagues. There was always another interesting idea to huddle over or an animal breaking one rule of psychology or another.

His students' recollections of him, from this time, balance between affection and amusement. He was thinner now, slighter, more burned out in appearance. A seemingly fragile man, constantly hurrying through the halls of the primate lab, almost tilted forward in his hurry. "Harry Harlow did not just walk," former graduate student John Gluck wrote. "He walked bent precariously at the waist in a manner which required that his feet shuffle rapidly forward to prevent him from tipping over."

When Bob Zimmermann came to the Wisconsin lab, he was sent directly into a conference room to meet his new professor: "There were four or five people. I'd never seen them before. And one was a guy with a crew cut. Another was a distinguished guy in a suit. And then there was this little guy with glasses, wearing a sloppy shirt and he had on these, I forget what you used to call them, but they were denim elastic-top kind of pants. This last guy came walking out with a cigarette dripping, just hanging out of his mouth and his hand comes out and he says, 'I'm Harry Harlow.'"

Harry had always had the potential to be an eccentric. He'd been an oddball kind of child, dreaming of trains to the moon instead of settling into the solid decent Iowa culture. A quirky kind of graduate student, with his poetry-sprinkled notes and wry approach to doing what he was told. Everything that had happened to him at Wisconsin—from his zoo-based research to his fallen first marriage—had conspired to nourish the offbeat parts of his nature. He was in his late forties now and he no longer cared about being someone else's image of a rising psychology star. He might have hidden his poems while at Stanford but now he left them on his students' desks at night.

Harry had never cared much about status symbols; clothes and cars and houses didn't impress him. He liked to be paid well. He just wasn't interested in wearing his money. Peggy, who liked to hoard for the future, encouraged that indifference. Helen LeRoy, who was both assistant and friend, recalls driving with Harry to pick up a visiting Soviet scientist at an elegant lakefront hotel. They swept up the circular drive in the Harlows' tan-and-cream Chevrolet. It was their one car, it was almost ten years old, it was covered with dings and rust spots. LeRoy remembers thinking, "At least he won't think we're rich American capitalists."

Harry was not at all self-conscious about the car. Or the clothes. He was so consumed by the ideas, bubbling inside, that he rarely remembered exterior appearance. Or even what he was wearing at a given moment. He left a trail of hats, coats, and scarves across the country every time he traveled for business; his files are full of query letters to friends and hotels, wondering whether they had found his garments. LeRoy sometimes worried for him. It bothered her that Harry looked increasingly as if he had dressed himself at a thrift shop. This came home to her on a day of errand-running. Harry liked to visit his bank, First Wisconsin, in the morning when it routinely put out coffee and donuts for customers. One day, they were standing together by the coffee table, talking, when a teller called out, "Dr. Harlow? Could I ask you a question?"

As Harry walked over, a woman standing next to Helen whispered, "Is that Harry Harlow?" She nodded.

The woman gestured toward Harry's thready overcoat, saying, "I thought he was someone who came off the street for coffee."

Still, LeRoy laughs remembering. Harry thought it was funny. He could never resist a joke on himself. He didn't mind being odd as long as he made an impression. "He'd say, if you want them to remember you, make them laugh," says a former graduate student, Lorna Smith Benjamin. That same casual indifference toward impressing others certainly characterized his old laboratory with its shaggy yard and chipped walls. He was proud of it, fond of it, he might have kept that laboratory indefinitely if the University of Wisconsin hadn't almost forced him into something better. As far as Harry could tell, the change wasn't a vote of confidence in his research; it was a vote against the appearance of his laboratory. The nearby engineering department had complained about the eyesore in its backyard. The university's president had toured the lab and informed Harry that he looked forward to tearing the dump down. "I was shocked," Harry wrote. "This was my bride and all brides are beautiful." Still, he was happy to put in a request for a larger building and remodeling money and, in the end, the university did give him a new building—this time an abandoned cheese factory—and refurbishing money as well. His second laboratory lacked a yard but it was at least three times the size of the first.

At his new building, finished in the early 1950s, Harry put together the kind of creative lab family that suited him. He didn't care whether staff and students came from a status school or had a perfect academic track record. He created a laboratory filled with people who were interested in the work and people who interested him. He accumulated students recommended by friends, students he liked, and students he thought were just plain smart.

Gerald McClearn, now an acclaimed behavioral geneticist at Penn State, remembers that he went to Wisconsin because Harry happened to give a speech at his small undergraduate college. McClearn's major professor at Allegheny College marched him up to Harry afterwards and said, "Harry Harlow, you don't know me, but if you don't take this boy to be a grad student, you're missing a big bet." "[Harry] looked amused, asked me a few questions, and after five minutes, he said, 'Okay, you'll do. Call and tell them you're in.'" McClearn laughs: "I hadn't even applied."

Once there, "You were essentially encouraged to do whatever you wanted to do," says John Gluck, now a psychology professor at the University of New Mexico. "He didn't give you specific work to do. The lab had tremendous resources and you were expected to make use of them." If he thought you were any good, he gave you enormous room to experiment—so much room that he sometimes lost track of what students working in the lab were up to. Another student remembers that after he had completed his Ph.D. dissertation, he handed the thickly stacked papers to Harry. The professor halted, studied the document at hand and said, "Okay, now I'll find out what you've been doing all this time."

Another of Harry's research assistants, Marvin Levine, had come from B. F. Skinner's lab at Harvard. Skinner was working only with pigeons when Levine was there. The great man rarely showed up in the lab; instead, he gave the lab managers instructions to be carried out in his absence. Every graduate student knew those instructions had to be carried out with military precision. Skinner liked his lab formal in manner and orderly in operation. He had even written guidelines for how to take pigeons out of a cage and how to return them. So that Levine also remembers the almost cold-water shock of coming to Harry's lab, where he "would give you as much responsibility as you could take."

The first impression could be deceptively chaotic. Over the years, Harry had clinical psychology students, hard-line experimentalists, and students interested in learning, fascinated by emotions, wondering about relationships between animals. He allowed students who had no idea what they were interested in at all. The permanent staff of the primate laboratory was composed of people holding degrees in psychology, economics, literature, music, or what John Gluck liked to call "diplomas in curiosity."

Harry supported students through the fine ideas and the fanciful ones and the failed ones. One of his most inventive graduate students was a New York–raised psychologist named Leonard Rosenblum. While at the Harlow lab, Rosenblum invented a mechanical head to scare monkeys. He was interested in exploring animals' fear responses. "I decided to make a threat face. In order to do it, I sculpted a head out of balsa wood with a hinged jaw and hidden teeth that would be exposed when the jaw opened. Another thing, the ears would flap. So, when I pressed a switch, the jaw opened, the ears flapped back."

Rosenblum put this theoretically terrifying face onto wheels. He wanted it to roll on a little track, something like a toy train. The track ran into a playroom where he was already working with a group of monkeys. But when he rolled out the head, the monkeys—long-used to new experiments by enthusiastic students—yawned. "Harry's experienced monkeys really didn't pay much attention the first couple times," Rosenblum continues. He decided he needed more drama. "So I attached a buzzer to catch their attention. I attached it but I didn't insulate the lead wire. I decided to show Dr. Harlow to get his approval. That wasn't easy to get from Harry. Not overtly, not to your face. I cajoled him into seeing this."

They stood together in the playroom as the head rolled out on the track. As it moved forward, the exposed leads struck the track. Sparks shot up. The circuits blew. The room went dark. Rosenblum braced himself for an explosion from his professor. Quietly, the door leading out of the room opened. Harry stood silhouetted against the light. "Very impressive, Rosenblum," he said.

"That was it. He left, he never mentioned it again and I was never in any way punished for it." Even now, Rosenblum can't help grinning at the memory. This is not to say that Harry would allow a faulty idea to continue indefinitely, and Rosenblum remembers that as well. In his first year at Harry's lab, he had decided to test the social activity of monkeys in a maze. Rosenblum wanted it to be a memorable maze. And it was. It filled an entire room; it crowded against the walls. The monkeys drifted around in the maze, idly curious. But there wasn't any notable social difference to measure. "I was months in it and it wasn't working but I was doggedly at it. Harry didn't say a thing.

"And then one day he came up and said, 'Rosenblum, take a sledgehammer to that thing and get rid of it.'"

"I said, 'Get rid of it?' Rosenblum's voice rises as he recalls his shock.

"He said, 'Get rid of it. You've got to know when to quit.'"

And again, that was it. "He never held it against me and he was right—you do have to know when to quit. If you cling to errors, you never learn the right way. It was a very difficult lesson. I was embarrassed and ashamed. But he didn't see it that way. It was that it was over. It was a hard lesson and an important one and I've held onto it." Rosenblum went on to a heralded career at the State University of New York, directing a behavioral primate laboratory in Brooklyn, and exploring the social development of children.

"Harry never punished you for trying," Rosenblum continues. "He was at his best, most sympathetic, when you were down. He never turned his back on people who screwed up." Harry would even rescue people. Another graduate student, Kenneth Schiltz, had been unable to survive the statistics gauntlet of the psychology department. When Schiltz dropped out, Harry almost immediately gave him a lab job. As many students remember, including Rosenblum, Kenny Schiltz took over the nurturing and the handholding duties at the lab. It might have been difficult to wring overt praise from Harry, but Schiltz loved to tell people that they were doing well. "This lovely man, a defunct ex-grad student of Harry's, stayed with him for years, acting as a drinking companion, a sort of lab manager, researcher, and older brother to the grad students," Rosenblum says.

And brother is a good word for the relationship because, for Harry, the lab was family. Because it was often his first family, he came to expect his students to see it that way, too. If grad students had children, they often brought them to the lab. Many have memories of babies bundled into blankets and sleeping in offices or the break room. "The first year I was there, I said I can't be at a meeting, I'm going home for Thanksgiving. And Harry just looked at me and said, 'You're going to have to give all that up,'" recalls Lorna Smith Benjamin, now at the University of Utah.

Harry was at his laboratory so much that people started to wonder whether he ever went home. "I mean this kindly; Harry had idiosyncrasies," said Richard Wolf, a professor of physiology at the University of Wisconsin medical school at the time. Wolf collaborated with Harry on several projects. "Harry believed that you had to be at the lab seven days a week. One Monday morning he came by my lab, and said, 'I didn't see you yesterday.'"

Harry took pride in being the man who was always there. He could even be competitive about putting in more time than anyone else. Gene "Jim" Sackett, one of Harry's most valued postdoctoral researchers, used to set his alarm so that he would be the first one at the building. "Harry and I vied for who would open the primate lab. We were both early birds. I would sometimes come in at 5 or 5:30

A.M., and he would come in at 5 or 5:30 A.M., and it would be almost a game, without talking about it, who's going to open the door; because whoever got there first opened the door."

"You're there at night, he's there at night," Gluck once wrote in a testimonial to his former professor.

> He'd roam the halls at night leaving love poetry on graduate students' desks, checking doors, looking for someone to talk to. He'd invite any comers for a cup of the jet black coffee smoking in the urn by the surgery prep room. He'd invite you in to read a manuscript, where inserted sentences meandered like ant trails around margins and corners. If you couldn't keep up, if he lost interest, he'd just put his head down on his desk, pillowed by his crossed hands.
>
> Questions raced across the mind. Was he asleep? Did I say something that stupid? Do I go on talking? Should I poke him gently to see that he is all right? After a while, you got used to Harry's head going down. It was a message. It meant I'm done with this—go back to work.

If that didn't work, he could be even more direct. He'd stick his head into the student break room and interrupt a card game, with a single question. "Making headway?" his tone just sarcastic enough to empty the room. "Do you have an office?" he snapped to one student, who nodded dumbly. "Then go use it."

Mostly, though, Harry Harlow would just be there, talking to students, the coffee cooling at his elbow and the cigarettes, burning, unnoticed between his fingers, smoke coiling like dreams, the ash glowing redder and redder as it neared his skin. His grad students used to stare, mesmerized, at the imminent collision of hot ash and bare fingers. As with his unfashionable and battered car, his forgotten hats and scarves, he was simply more interested in the conversation or the idea than whatever was in his hand at the moment. Helen LeRoy once tracked Harry down at a scientific meeting by following a trail of partially smoked and smoldering cigarettes. At least he always dropped them into ashtrays and wastebaskets. One morning, he did that at the lab and the basket burst into flames. LeRoy simply poured her coffee into the fire and put it out. Harry was so busy talking that he never noticed the blaze.

He seemed to exist on coffee, smoke, and alcohol. He was drinking more than ever by now, not just when he went home but with friends, with students, at local bars and professional meetings. Often, he brought a bottle of liquor—bourbon or scotch—to stock his hotel rooms when he traveled. Of course, in academic psychology, in those days, liquor flowed the way wine does today, only more so. "People don't party now the way they used to," said William Verplanck, the former Hullian scholar, sadly. "Until the early 1960s, the professional meetings were awash with alcohol. Now if there's a meeting of psychologists, there's a room reserved where people go up to a cash bar." His voice crackled with disgust and regret for lost times.

"During the thirties, forties, fifties, a university would have a hotel room where people would gather and bring their own bottles or steal somebody else's. There'd be lots of people sitting on the floor in the hall drinking, and talking, and occasionally going back in. Everyone spent some time loaded. Now it's so bloody formal, it's like a meeting of CPAs. We drink less and we communicate less." Plenty of old colleagues share Verplanck's sense of leaving behind a more exciting time in psychology. "One of my friends insists that it's a waste to die with all your organs intact," says an old friend of Harry Harlow's, his voice also rich with nostalgia. "We're all so cautious now about what we say and think and do. It used to be a lot more interesting—and more fun. It was a spicier time."

Harry's habits suited that time of smoke and drink and rapid-fire conversation perfectly. In some ways, he surpassed it. Alcohol was becoming an integral part of his life. He even wrote odes to it: *Clover club's a nice girl / Vodka is a shrew / Corn whiskey is the old love / Scotch whiskey is the new.* When he was burned out at the lab, he liked to take an early evening walk down to a nearby tavern. "You knew you'd arrived when Harry asked you to go for a walk," Zimmermann said. "Harry'd come in, pick someone out, and say, 'Let's go for a walk.' You'd usually go for a beer, sit down, and sometimes he'd talk and sometimes he'd never say a word. I remember I got there in July and September was my first walk. That was a thrill. We went to this bar, and ordered a beer, and he said, 'Got a dollar?'"

It still makes Zimmermann laugh.

"He called me Jim, which surprised me because he mostly called people by their last names," recalls Jim Sackett, now a professor of psychology at the University of Washington in Seattle. Sackett occasionally worried that Harry Harlow would just burn up on alcohol fumes and psychology dreams. "He'd say, 'Jim, let's go to the corner.' So we'd go the bar on the corner. By the time we'd left, he'd had three, four drinks. I'd had one. And I'd drive him home and think, 'God, he's gonna die, he's just out of it. He's gonna be dead.' The next morning he'd be there at 5 A.M. He'd be there and he'd be writing."

And it was during those hyperactive, free-form, sleep-deprived, alcohol-inspired days that Harry Harlow first started thinking about the nature of love. That he got there at all can seem improbable. You could make a case that this was the least likely laboratory to take on the cause of love, this outpost of Wisconsin psychology where work came first and family last. You could argue that Harry would be the most unlikely of champions, a man whose world turned on monkeys and primate labs and graduate students arguing their theories over coffee and cards and nights at the corner bar. He was a father who left the house before his children were awake, a man trailing a failed marriage behind him. He was sarcastic, edgy, and completely opposed to sentiment.

You could also make the case that Harry Harlow was absolutely perfect for the job—a man objective enough about love to see it as the stuff of science. A psychologist who still allowed for all the possibilities. He was a scientist with a love of the creative, a professor willing to give his students every chance to let their ideas take flight; a man who still wrote poetry in the night, who supported young researchers who didn't agree with him. He was a man who could laugh at his own mistakes. The hard times had helped make him a psychologist who didn't worry about fitting in or making the right impression. They hadn't stopped him from being a dreamer, though, or a lover of lost causes or a man who could look at a lab full of burly, quarrelsome rhesus macaques and start thinking about the importance of mothers and the needs of children.

SIX

The Perfect Mother

One cannot ever really give back to a child the love and attention he needed and did not receive when he was small.

John Bowlby,
***Can I Leave My Baby?*, 1958**

STILL, HARRY DID NOT STEP directly into love; there was no triumphant flourish of research trumpets. In 1955, he had to tackle a different problem, more pragmatic, more urgent. It had to do with importing monkeys: He was beginning to hate that process. The animals were hard to find. They were expensive. They were often in terrible shape. Monkeys routinely turned up starving, battered in passage, seething with "ghastly diseases." The hot-tempered, tropical viruses spread easily. The incoming macaques infected their cagemates. Playmates sickened alongside monkey playmates. Macaque mothers passed diseases to their infants. A laboratory with a new shipment of monkeys could more easily resemble a hospital than a research laboratory.

Harry began to ponder raising his own animals. It was this decision that would, indirectly, lead him into the science of affection. When it did—when he first started wondering whether you can raise a healthy child, even a monkey child, without love—the people working with him would think he'd gone crazy. Of course, they were used to Harry Harlow's crazy ideas. Starting a breeding colony in Madison, Wisconsin, struck plenty of people as evidence enough of lunacy. The Midwestern climate, almost the polar opposite of the balmy seasons of India, seemed an unlikely place to start raising tropical species. But Harry had been accommodating monkeys for many winters. He figured that they'd just continue bundling the monkeys inside. That would keep the colony small, only what he could house indoors. He could live with that.

There was another, bigger, challenge. No one really knew how to do what he wanted. There were no self-sustaining colonies of monkeys in the United States. The domestic breeding of primates was a brand new, barely simmering idea. Other people were talking about it; indeed, researchers from California to Connecticut were equally frustrated. But no one had any experience at breeding monkeys on the scale Harry imagined. Only a few American scientists had even tried hand-raising the animals in any systematic way and that had been on a monkey-by-monkey kind of scale. Did this faze Harry? Not really. Once you've built a laboratory out of a box factory, starting a breeding colony from scratch just isn't that big a deal.

Still, he first consulted with his friends at Wisconsin. Harry and his university colleagues decided to approach the problem like the scientists they were. What does one feed a baby monkey? William Stone, from the university's biochemistry department, spent countless hours testing formulas. As he remarked years later, "I can still smell the monkeys as I recall sleeping at the primate lab on a four-hour schedule" to try out different recipes on the baby monkeys. Stone eventually had so much data that he published a paper on the immune effects of feeding cattle serum to newborn monkeys. He began with a baby formula of sugar, evaporated milk, and water. He recruited students to hold doll-sized bottles to feed the monkeys. Every bottle was sterilized. The monkeys received vitamins every day. Their daily doses included iron extracts, penicillin and other antibiotics, glucose, and "constant, tender, loving care." The baby mon-

Psychologist B.F. Skinner of Harvard University, perhaps the most famous advocate of conditioned behavior in both animals and humans. *Photo courtesy of the Archives of the History of American Psychology*

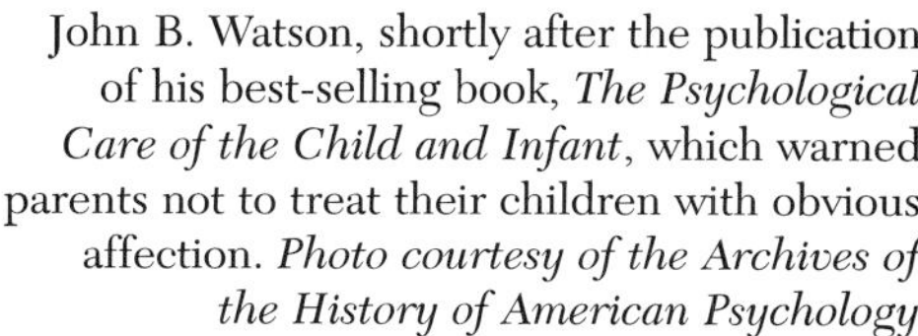

John B. Watson, shortly after the publication of his best-selling book, *The Psychological Care of the Child and Infant*, which warned parents not to treat their children with obvious affection. *Photo courtesy of the Archives of the History of American Psychology*

Clara Mears and Harry Harlow in 1931, in the happy year before their first marriage to each other. *Photo courtesy of Robert Israel*

Clara and Harry Harlow, and their sons, Robert, age 5, and Richard, age 2, in 1944, two years before the couple divorced and Clara and the boys left Madison. *Photo courtesy of Robert Israel*

Harry and son Robert, age 6, during one of their weekend visits to the primate laboratory, in a sideyard along the railroad tracks. *Photo courtesy of the Harlow Primate Laboratory, University of Wisconsin-Madison*

The former Wisconsin psychology department building—now demolished—which was given the nickname Goon Park because of its street address. *Photo courtesy of the University of Wisconsin Archives*

Peggy and Harry Harlow, shortly after their 1948 marriage, visiting with Harry's first son, Robert. *Photo courtesy of the Harlow Primate Laboratory, University of Wisconsin-Madison*

Harry Harlow's first primate laboratory, which he converted out of an abandoned box factory building. *Photo courtesy of the Harlow Primate Laboratory, University of Wisconsin-Madison*

Three young rhesus macaques puzzle their way toward opening a lock during a curiosity experiment at the Harlow laboratory. *Photo courtesy of the Harlow Primate Laboratory, University of Wisconsin-Madison*

Robert Zimmermann, a Harlow graduate student who worked closely with the mother love studies, takes a moment to relax at the laboratory. *Photo courtesy of Robert Zimmermann*

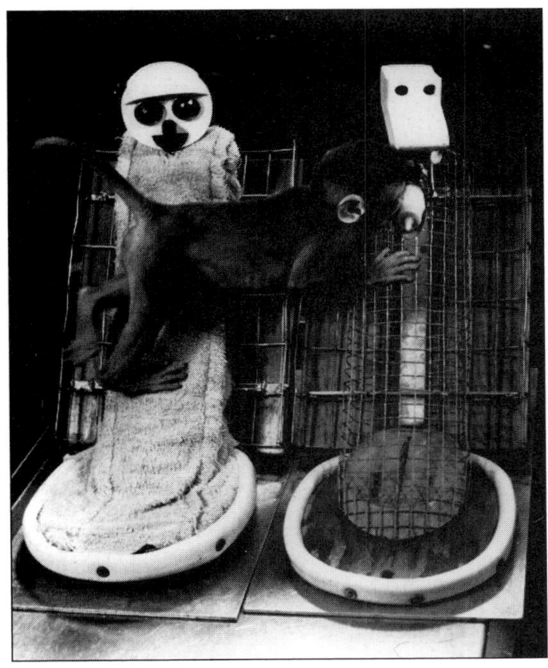

A baby monkey keeps a possessive grip on his beloved cloth mother while reaching over to wire mother to be fed. *Photo courtesy of the Harlow Primate Laboratory, University of Wisconsin-Madison*

A young rhesus monkey scoots back to his mother as a scientist approaches. *Photo courtesy of the Harlow Primate Laboratory, University of Wisconsin-Madison*

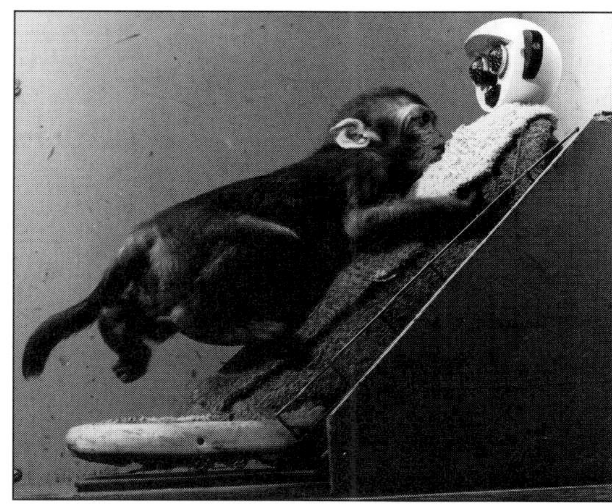

A baby monkey, startled during an experiment, leaps for his cloth mother, who represents security and comfort. *Photo courtesy of the Harlow Primate Laboratory, University of Wisconsin-Madison*

The cloth and wire mothers, side by side, from the original Harlow tests of the importance of contact comfort. *Photo courtesy of the Harlow Primate Laboratory, University of Wisconsin-Madison*

A baby rhesus macaque, in a new and strange room, with no mother nearby, gives way to fear and loneliness. *Photo courtesy of the Harlow Primate Laboratory, University of Wisconsin-Madison*

A wind-up toy drummer bear was one of the devices used to test the fear in young monkeys and whether a mother provided a sense of security. *Photo courtesy of the Harlow Primate Laboratory, University of Wisconsin-Madison*

In this 1958 publicity photo, Harry surveys one of his most famous results, the union of cloth mother and a baby monkey in her care. *Photo courtesy of the University of Wisconsin Archives*

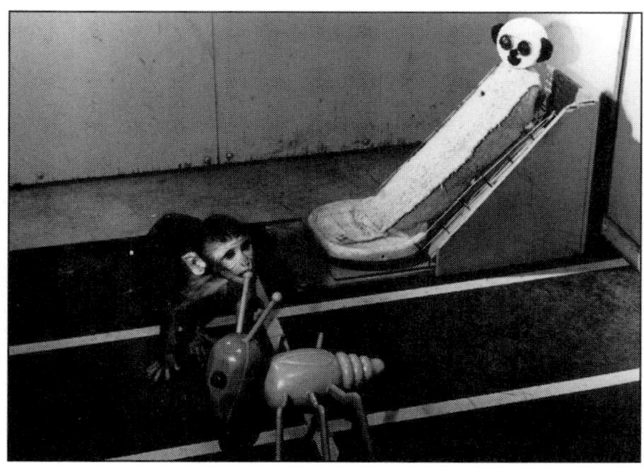

A baby monkey, in the comforting presence of his cloth mother, decides to tackle a previously frightening toy insect. *Photo courtesy of the Harlow Primate Laboratory, University of Wisconsin-Madison*

Harry, taking a rare moment to relax, during the late 1960s. *Photo courtesy of the Harlow Primate Laboratory, University of Wisconsin-Madison*

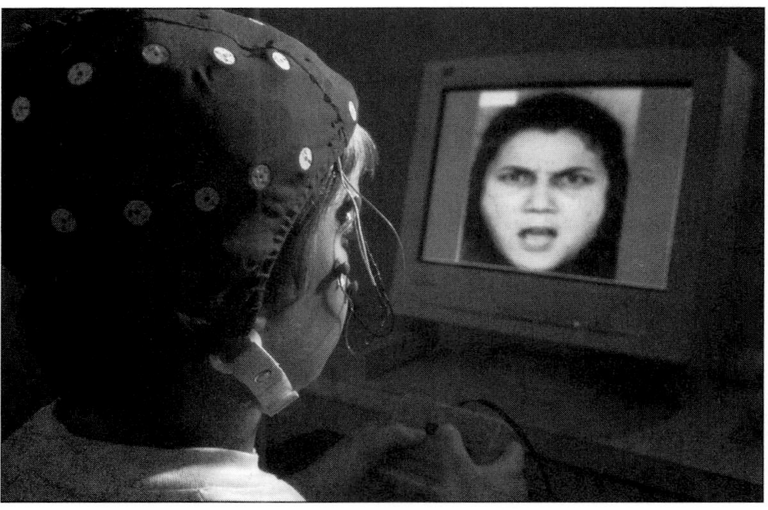

A young boy evaluates an angry face during a recent study of children's emotional relationships at the University of Wisconsin Department of Psychology. *Photo courtesy of University of Wisconsin News Services*

keys were washed, weighed, and watched over constantly. As the monkeys grew older, lab caretakers mixed fresh fruit and bread into their diet. And always, always, the caretakers kept the animals apart from each other. Every monkey in a separate cage. Every baby taken from his mother, which is why someone needed to hold those baby bottles. Harry wanted no chances taken on the spread of those ghastly diseases. Everything was polished and cleaned and disinfected and wiped to a glittering cleanliness.

There was a model for such practices in human medicine, in the frantic efforts of early pediatricians to control disease in orphanages and hospitals. The Wisconsin researchers mimicked perfectly, had they realized it, the very hospital policies that Harry Bakwin had been so furiously trying to undo in the 1940s. Harlow and his colleagues were inadvertently recreating those isolationist pediatric wards.

By the end of 1956, the lab managers had taken more than sixty baby monkeys away from their mothers, tucked them into a neatly kept nursery, usually within six to twelve hours after a monkey's birth. Lab staffers fed the infant animals meticulously, every two hours, with the carefully researched formula from the tiny dolls' bottles. And the monkeys looked good. The little animals gained weight on that formula. They were bigger than usual, heftier and healthier looking. And they were purified of infection, "disease-free without any doubt," wrote Harry. But their appearance, he added, turned out to be deceptive: "In many other ways they were not free at all."

The monkeys seemed dumbfounded by loneliness. They would sit and rock, stare into space, suck their thumbs. When the monkeys were older and the scientists tried to bring them together for breeding, the animals backed away. They might stare at each other. They might even make a few tentative gestures, as if each primate vaguely wished to encourage something. But the nursery-raised monkeys had no idea what to do with each other. They seemed startled by the appearance of other animals, intimidated by the sight of such odd, furry strangers. The monkeys were so unnerved by each other that many of them would simply stare at the floor of the cage, refusing to look up. "We had created a brooding, not a breeding colony," Harry once commented.

How could the monkeys look so healthy and yet be so completely unhealthy in their behavior? The researchers had a growing colony of sturdy, bright-eyed, bizarre animals in their cages. Not all the animals were so unstable. But enough were to keep the researchers up at night. Harry was driven to making lists of possibilities. What was he doing wrong? Could it be the light cycle—was the lab not dark enough at night? The antibiotics? Perhaps the medicines were skewing normal development. The formula? It might be that evaporated milk wasn't such a good thing. Maybe the baby monkeys were being given too much sugar—or not enough.

Harry and his students and colleagues talked it over as the coffee steamed, the bridge cards shuffled, and the nights burned away in the lab. Harry's research crew was still growing and, on the recommendation of his old professor, Calvin Stone, he'd brought another Stanford graduate into his lab. The latest young psychologist to venture into the box factory was named William Mason. His Stanford Ph.D. barely off the presses, Mason found himself immediately plunged into the problem of the not-quite-right baby monkeys.

Shortly after arriving, Bill Mason was put in charge of raising six newborn animals. These were all lab-made orphans: taken away from their mothers some two hours after birth. In Harry's lab, the monkeys were often given names instead of the numbers that are standard in primate labs today. The oldest of Mason's orphans was Millstone, named by a lab tech because the little monkey was such a noisy, clingy pest. The other five infants also joined the Stone family: Grindstone, Rhinestone, Loadstone, Brimstone, and Earthstone. A research assistant at the lab, Nancy Blazek, had feeding duties. Exhausted by the two-hour schedule, she took to bringing the little monkeys home with her for their nighttime bottles.

Mason and Blazek spent hours with those monkeys, and they got to know them well. They wanted the babies to grow strong and healthy. Mason planned to continue some of the earlier studies on curiosity. Harry had established that monkeys were naturally curious, and Mason wondered how early that trait showed up. When did monkeys start to wonder about the world around them? Were they born asking questions or did they pick it up later? When it came to puzzles, at least, Mason and Blazek found that the Stone babies were naturals. As soon as they were coordinated enough to work a puzzle, the little creatures were busy trying to solve it. The results reinforced a strong suspicion that curiosity was fundamental to the way these small primates approached the world.

Something else about the Stone monkeys caught the lab workers' attention. The researchers had been lining the cages with cloth diapers to provide a little softness and warmth against the floor. All the little monkeys, including those in the Stone family, were absolutely, fanatically attached to those diapers; they not only hugged the diapers fiercely but also wrapped themselves in the white cloth, clutched at it desperately if someone picked them up. Around the lab, an observer might be struck by the appearance of baby monkeys in transit, cloth streaming out behind them like kite tails.

There was already a hint about this cloth-obsession from the nineteenth century, in the diaries of a British naturalist named Alfred Russel Wallace. The adventurous Wallace is best remembered now because he so nearly published a theory of evolution before Charles Darwin. As with Darwin, it was traveling that made the theory come to life. Exploring the oddly different and beautifully adapted species of each country also made Wallace think about the way nature tucks us into our niches. During a visit to Indonesia, Wallace had been given an orphaned baby orangutan. He wrote in his journal that the little animal seemed to constantly reach for and cuddle soft material, including (painfully) Wallace's beard. Trying to help the baby and himself, Wallace made what he called a "stuffed mother" out of a roll of buffalo skin. He noted that the little ape clung happily to the fat, fuzzy roll, no longer interested in random beards and shirts.

And there was another, more recent, clue from a Yale University researcher famed for her meticulous comparisons of monkey and human anatomy. To make a detailed analysis, Gertrude Van Wagenen had needed a reliable supply of monkeys. She'd created a small nursery and written an insightful chapter on her technique for raising baby monkeys. Van Wagenen had found that her nursery-raised monkeys bonded almost compulsively to the soft blankets lining their baskets. She described their tight clutch as emotional dependency, and noted that if the infant monkeys couldn't cuddle, some of them didn't even develop proper feeding responses. "You know of the debt I owe to you for the creation of the rhesus baby in the basket," Harry wrote to her, late in his career. "The early research, which I conducted according to your directions, started me off in the field of primate affection."

And, indeed, the psychologists in his primate lab were beginning to wonder whether there was a real message in the behavior of their baby monkeys. Perhaps the small animals had something to tell them about the needs of children; after all, it wasn't monkeys alone who clung to soft cloth. Orangutans, too, did it, and other labs reported that baby chimpanzees desperately hugged blankets. Nancy Blazek and Bill Mason, watching their monkeys cling to the cloth, started wondering about that need to hold on. They had that other primate to consider in this idea, too. All of them knew that human babies, left alone in a crib, also clutched their quilts and pillows and fluffy stuffed toys. But what did that mean? What was the message in the apparent magic of cloth?

Mason suggested to Harry that they run a test. He was thinking of a simple comparison between, say, a fat bundle of cloth and something hard—wood or wire. The researchers could see what the babies preferred—whether it was just the need to hold on to something, anything, or whether there was something especially meaningful about a soft touch.

And the idea just clicked for Harry. He liked it immediately. He also thought there might be something even bigger lurking there. Perhaps the differences between cloth and wood touched on part of the underlying question only. After all, babies don't prefer to hold on to pieces of cloth over all else. They hold onto them when there's no human—or monkey—available for cuddling. The soft bits of cloth might be a substitute for something that mothers do that's missing. Today, of course, we would include fathers, but this *was* the 1950s; and at that time, in science and society alike, it was mother who represented what parents had to give a child.

So, if Harry was right, if they were looking at an odd, pathetic kind of mother substitute in these blankets, they were also looking at raising a revolution in psychology. If the baby monkeys were telling them that there was something critical in being touched, in being held and in holding back, then they could start rewriting the psychology books. And the first new sentence in that book might say that mothers themselves—with their soft arms and inclination to hold a baby close—were desperately important; and if that was right, the Watsonian, Skinnerian, Hullian view of the world could be nothing less than wrong.

◦ ◦ ◦

Harry used to say that the idea for a lab-built mother occurred to him on a Northwest Airlines flight between Detroit and Madison as he looked out at the puffy, deceptively soft clouds billowing on the other side of the glass. "As I turned to look out the window, I suddenly saw a vision of the cloth surrogate mother sitting beside me," he wrote. A lab-created doll of a mother, as deceptively soft as those floating clouds, could be used to see what a baby really wanted. It would be a comparison, as Bill Mason suggested, but it would be a comparison that used a mother figure, one that looked obviously enough like a mother that anyone could see that this was not only about monkeys.

Harry Harlow had encouraged the students and employees in his lab to think for themselves, and they didn't hesitate in this instance. They thought he was wrong. As far as Harry could tell, his students thought that their major professor had left his head in the clouds: "My enthusiastic descent upon the laboratory was met by skepticism or lack of interest from one graduate student after another." He was finally able, he said, to convince one of his newer graduate students, Robert Zimmermann, to give it a try. "But I'll tell you one thing about those damn airplane rides when we were on the surrogate project," says Zimmermann, now retired in Lansing, Michigan. "Every time Harry would fly somewhere—and he went away every week because he was on all kinds of committees—he'd run into some shrink or somebody and he would come back with some new idea about what we should be doing with the surrogate. He'd always wonder when he came back, 'Why don't we have this? So and so said we should have rocking, why don't we have rocking?'"

Zimmermann is laughing when he tells this story. He agrees that Harry was right about one thing—most of the students fled from getting sucked into a project as mushy, as un-Wisconsin, as mother love:

> In all honesty, nobody, no grad student wanted to touch the mother surrogate project with a ten-foot pole. This was Wisconsin, and Harry could be of some help, but you had to get your thesis or dissertation past a committee, and to talk about love at the University of Wisconsin, where everything was numbers and statistics . . . I think the first assumption was that if you took that one you'd never graduate.
>
> Well, I was already working with neonatal learning and nothing much was being done with the babies in the first ninety days of life, before they were ready for those experiments, and I thought, well, I have an investment in these monkeys, so I made a deal with Harry. I would be the ramrod for the mother surrogate project if he would let me have the baby monkeys for my dissertation. And he thought that would be a fair trade.

The airplane birth of the surrogate mother—the way Harry would tell it, full of drama and imagery—says a lot about Harry's vision for the project. Here was science at its most provocative—mother love at a time when British psychiatrist John Bowlby could barely persuade his colleagues to join the words *mother* and *love* together. Here, also, was science with real potential to make a difference, to make people see families and relationships in a different way, a closer way. The first challenge would be getting anyone to take it seriously.

That was going to take both solid research and, Harry suspected, all the skills at making an idea compelling that he had acquired over the years, all the unflinching stubbornness he had learned while he wangled a laboratory from the University of Wisconsin and persuaded his colleagues that maybe, just maybe, monkeys were smarter than scientists had thought. If he wanted an attentive audience—and, oh, he really did—the surrogate mother was going to be a Harry Harlow production.

His newly minted Stanford researcher, Bill Mason, was stunned by how rapidly his small, neat idea became a showstopper. "I didn't see it as a breakthrough or something really sensational," Mason says. "It was a kind of demonstration with a foregone conclusion." There was Wallace, after all, and there was Van Wagenen; everyone in the lab expected the monkeys to prefer the cloth. They worked out a kind of trial balloon. Zimmermann teamed with another graduate student, Lorna Smith, and the two of them did a simple first test with two baby monkeys. Both the little animals flatly rejected a wire object in favor of a cloth bundle. "It was unbelievably clear, amazing," Zimmermann says, and suddenly the lab crew began to consider the possibility that Harry Harlow was going to pull this love stuff off after all.

From that first experiment, Harry wanted everything nailed, every detail noted, every possible criticism identified and answered. He insisted on two observers for every experiment with the little animals, one student double-checking the other. He devised careful charts to score the monkeys' behavior. Harry and his students filmed the experiments and then spent hours scrutinizing each frame, right down to the clasp of the fingers on the cloth. "He was concerned it would be rejected out of hand if we didn't nail it to the floor," Zimmermann says.

Mason still remembers, with admiration, Harry's skill at taking a long-dismissed idea—that mother love was a crucial part of a child's development—and persuading his colleagues to listen to him. "The dominant position was that babies didn't love their mothers or need them, that the only relationship was based on being fed," Mason says. "It sounds silly now but that's what people thought. Harlow sensed people were beginning to ask questions. And it was damn right to ask questions because the dominant position wasn't true. These are facts—monkeys don't just explore for food, they do it because they are curious, they have a drive to know. And they are social and they need to interact. Harlow had a great sense of when he could get away with challenging the field. If he had misjudged that—if he had been younger, less skillful—it would have been a disaster. People would have laughed."

Skillful or not, there was no doubt that Harry was yet again on the wrong side of behaviorist psychology. B. F. Skinner was now experimenting with boxes in which to raise young children. Skinner had built the first demonstration model for his younger daughter, Debbie. It was a crib-sized "living space"—a baby-tender, Skinner called it, with sound-absorbing walls, a large window, and a canvas floor. The air in the box was filtered and humidified and the baby stayed so clean in there that Skinner said she needed a bath only twice a week. The partial soundproofing meant that the child was undisturbed by doorbells and ringing phones—or the voices of her parents and sister. Debbie came out for scheduled playtimes and meals: "One whole side of the compartment is safety glass, through which we all talk and gesture to her during the day. She greets us with a big smile when we look at her through the window," Skinner wrote in a letter to a friend, emphasizing the advantages of raising your baby in a box. He hoped that every mother would one day use a baby-tender. Skinner wrote once of being surprised when a pediatrician suggested that the box might be better used in hospitals where it could save nurses much work. It could save mothers work, too, Skinner replied. The doctor laughed. Mothers didn't care so much about the saved labor, he assured the psychologist. Mothers labored out of love.

"The universal reaction was, 'What is this love?'" recalls former grad student Leonard Rosenblum. "The only emotions studied in animals were negative—fear, loathing, pain. The idea that animals were motivated by love, what vague notion was this?" Rosenblum makes a dismissive gesture, indicating the disdain of the time. It has been a long time since he was a fledgling psychologist himself. He recently retired as director of a primate laboratory in Brooklyn, part of the State University of New York system. Today, Rosenblum is an internationally known expert in developmental biology, an angular man with bright blue eyes and a slightly shaggy silvery beard. He retains, though, the same intensity and lively humor and flair for a dramatic turn of phrase that he had as a student in Harry's laboratory.

"Remember," Rosenblum says, "that behaviorism's beginnings, with John Watson, suggested that it was a great thing to dig holes in the backyard and let your kids fall in and learn about life. So in psychology, love was smoke, mirrors, bullshit, and that was exactly what everyone was telling Harry." Of course, Harry was used to being told he was on the wrong side of an issue, the backside of the fence. He'd come to kind of enjoy needling the smugness of the mainstream position. He simply began assembling more evidence. Beyond that, he thought about how to make that evidence look really, really good. Bill Mason had proposed that they look at how monkeys might hold on to a bundle of cloth. And that was a beginning, said Harry, but they needed their surrogate to look like more than a bundle. It needed personality. It needed a head and a face. If monkeys were going to look at this substitute mother, it needed to look back at them. And it needed to look back at the human observers, too; it needed to mean something real to people. Harry wanted them all—not just psychologists but mothers and fathers and aunts and uncles and stepparents and grandparents—to think about connection and affection. He wanted them to believe that emotions and relationships were the proper purview of research.

Harry's students still sometimes argue about the decision to put a head on the cloth mother. Mason considers the head merely showmanship, unnecessary to testing monkeys, who, after all, would happily cuddle with a diaper. Others consider it strategy. One such former student, Steve Suomi, now head of primate behavioral research at the National Institutes of Health, still thinks of the head as a brilliant tactical move. "So it might not have been relevant to monkeys," Suomi says. "But it was to the outside world, because once people looked at the surrogate like a mother—made a connection to human mothers as well—then you could start talking about things like mother love."

And despite the fears by Mason, and even Bob Zimmermann, that the head was going to get them laughed out of psychology, Harry was absolutely determined. He had suddenly been given a first-class platform for his arguments. He'd finally—as Lewis Terman had predicted—been elected president of the American Psychological Association. And he was going use that position, he decided, to pound the podium and make his argument. He was absolutely sure of what he was going to argue. He had even thought of a title. He was going to call his talk "The Nature of Love." And as Zimmermann still remembers, "He came back into the laboratory and said to me, 'Bob, I have written one of the finest speeches ever delivered to the APA as president. Go get me the data.'"

So they put a plain wooden ball on top of the bundled body. Harry still wasn't satisfied. "It doesn't have a face," he said. By this time, Bob Zimmermann had fully taken over the construction project and he was willing, if he had to, to put a face on the surrogate mother. Harry had recruited Zimmermann, a tall, lanky man with dark hair trimmed into a ruthless crewcut, from Lehigh University in Pennsylvania. He was a promising and ambitious young psychologist. Zimmermann had received three offers from graduate schools, but Harry had written "the most beautiful letter" about Wisconsin and the land and the support the school could offer. After Zimmermann was properly seduced, he remembers, Harry wanted the letter back. It had worked so well that he wanted to try it on the next year's crop of recruits.

Zimmermann took the problem of the cloth mother's face to the lab's resident equipment-building genius, Art Schmidt. Schmidt was another typical Harry Harlow hire. He was a geography major and a hell of a handy guy. Even in his seventies, Schmidt remained lightly, toughly built. He still raced cars for fun. He had steady blue eyes and a slow smile. "I can build anything," he said simply. To help pay his way through school, Schmidt took a 90-cents-an-hour job at the primate lab, where he repaired cages and put up storm windows. When Schmidt graduated in 1953, Harry offered him a full-time job. Schmidt had gradually lost his enthusiasm for geography and he thought Harry would be a good boss. "He'd always go to bat for you if he thought you did a good job," he said. The new job partly required that he listen to the young researchers—who were all thumbs, according to Harry—and build their ideas into something functional.

Schmidt built the Butler box and long remembered Harry showing it off to visiting air force officials:

> It was to test how curious the monkeys were. There was a door they could push open and you could record how often they pushed it. I made a motor to open and close the door. The door would just wham down, and the monkeys were smart as hell, they'd just jump out of the way. And then they'd open it again. When these three or four colonels came, we all put on our clean lab coats, set up the box, and Harlow said, "This is Mr. Schmidt who built this and when Art Schmidt builds something it works." And then the door jammed. I was so embarrassed. But Harry just laughed. He said, "Except maybe today."

Oh, that head was a challenge, too. "First, it had to be designed to be pretty nondestructive," Zimmermann says. "Monkeys are very destructive creatures. And then it had to have eyes. Mothers have eyes, Harry said. So what are we going to do for eyes? So I go to these dolls' hospitals and stuff looking for eyes. If you've ever seen dolls' eyes, they're so fragile. And I said, 'Well, we need something that's a little stronger, that can take a knocking around.'

So this woman at the doll store says, 'Well, they're pretty expensive.' I said, 'Price is no object.'"

Zimmermann is grinning again as he tells this story, blue eyes crinkling at the corners. "So she says, 'You must work for the state.' "

They kept shopping. They didn't just want indestructible eyes. They wanted eyes that were also repulsive to monkeys. Harry had warned Schmidt and Zimmermann that the cloth mothers could not have faces that monkeys obviously found attractive. "Because then someone could say, hey, your experiment had nothing to do with touch or being held—it's just that it's an attractive stimulus," Zimmermann says. Critics might dismiss the cuddle effect and argue that the babies liked the way the face looked and that was why they clung to the softer body. "So we started fooling around with different configurations of faces, and then we would see how the monkeys reacted." They decided on bicycle reflectors for the eyes, which gave the face a bug-like stare. "Those are red bicycle reflectors. The mouth was green plastic, curved in a half moon smile. And the ears were a very hard black plastic that Art Schmidt had hanging around the lab. The nose was maple, painted black."

Schmidt and Zimmermann had even labored over what kind of wood to use for the heads. They'd tried pinewood balls, but the energetic monkeys chewed the soft wood into splinters. Zimmermann remembers complaining to Harry, "Dr. Harlow, the monkeys are destroying the heads. As fast as we make them, they're chewing 'em up."

And he remembers that Harry looked at him, completely deadpan, and replied, "Children have been destroying their parents for years." The researchers decided to use hardwood for the heads and settled on maple croquet balls, near rock-like in their construction.

Babies do chew on their parents, pull their hair, gnaw on their ears, drool on their shoulders, and throw up all over their shirts. Zimmermann points out that if you watch a baby monkey with its natural mother, the little guy will tug on fur, nibble on ears, yank and pull—all in affection. They'll do the same thing to a father monkey, given a chance. And this is not destruction at all; it's curiosity, touch, feel, and the infinite security of being held by someone who will put up with all that tugging and chewing. But monkeys and babies—as

Bowlby had been trying to say—indulge in those behaviors only with someone they love and trust.

One of the surrogate mothers in Harry's lab had a head but no face yet. The head was just a blank ball of wood because Schmidt and Zimmermann had not yet perfected the smiling mother's face. A baby monkey arrived a month early, so they put the animal in with the faceless cloth mother. "To the baby monkey this featureless face became beautiful and she frequently caressed it with hands and legs," Harry said. That lasted for about three months. "By the time the baby had reached ninety days, we had constructed an appropriate ornamental cloth-mother face, and we proudly mounted it on the surrogate's body. The baby took one look and screamed."

The little monkey huddled in the back of the cage, rocking in dismay. After several days, the infant solved the problem. She marched up and rotated the head 180 degrees so that the blank back of the ball faced forward. The scientists turned it back. She turned it again. They turned it. She turned it. "We could rotate the maternal face dozens of times and within an hour or so, the infant would turn it around 180 degrees." Within a week, the baby had resolved the problem entirely. She took the head off and rolled it into a corner of the cage and ignored it. And she was willing to repeat this; calmly, Harry said, and with infinite patience. He knew exactly what such behavior represented. Bowlby's theory predicted that one of the ways a baby bonds to a particular mother comes from its recognition of "the particular mother's face." It's that absolute sureness that this is *my* mother, that she's here, that makes everything all right.

The baby doesn't attach to just anyone, and John Bowlby, Harry Harlow, and a growing army of others were going to make that undeniably clear. There's an actual relationship here that matters; the baby recognizes this one special person as *the* one. Later studies at Wisconsin showed that monkeys definitely did not admire the face dreamed up by Zimmermann and built by Schmidt. They preferred a dog's face to the bug-eyed, green-smiled version of a mother. But for Harry, the antipathy also made a critical point. The infants might not like the mother's face, he said, but they loved the mother. She could have a blank face, a bug face, any face that they knew well—as long as she had mom's face. To a baby, mother's face is always beautiful, he said: "A mother's face that will stop a clock will not stop a baby."

The nature of love project was absolutely, beautifully straightforward in making that connection.

Art Schmidt built—as Bill Mason had first proposed—not one but two "surrogate mothers." The first was a cloth mother. She had that smiling face on a round head and a cylindrical body. The cloth mother was made from a block of wood, covered with sponge rubber, and sheathed in tan cotton terrycloth. A light bulb behind her back radiated heat. You could call her an ideal mother, Harry said, "soft, warm, and tender, a mother with infinite patience, a mother available twenty-four hours a day, a mother that never scolded her infant and never struck or hit her baby in error." The other mother had a squared, flattish face, two dark holes for eyes, and a frowning mouth. Beneath that scowling visage was another cylindrical body, also warmed by a light bulb, but this time made of wire mesh. It was perfect for climbing, but the wire mother had not a cuddly angle to her. She was metallic all the way through.

Eight baby monkeys went into the first surrogate mother study. Each went into a different cage. In every cage, two mothers awaited the little animal—a cloth mom and a wire mom. Would they prefer her with a warm, soft cloth body or a warm wire one? Oh, but Harry didn't want a question quite as simple as that. In four of the cages, cloth mom was also equipped to hold a bottle filled with milk. The other four had a "barren" cloth mother; wire mom held the bottle. So the experiment tested the prevailing theories of motherhood on two levels. If infants were indifferent to touch, they might be expected to shift equally between the two mothers. Unless the infant-mother relationship was based on food. If that were so, neither wire mom's stiff body nor cloth mom's pillowy one should make a difference. The babies should emphatically prefer the wire mom with the bottle to the cloth alternative. And if they did? Well, that would have been the end of Harry Harlow's research into the nature of love.

Instead, it was clear to Harry, hell, it was clear to everyone, that being fed formed no relationship at all for these baby monkeys. The mother love study suggested that the wire mother could have been dripping with milk, standing in puddles of the stuff, and yet the little monkeys wouldn't have cared for her. Cloth mom, on the other hand, was a magnet for a baby monkey.

In the published paper that followed, there are two small, neat, astonishingly clear graphs labeled "Fed on Cloth Mother" and "Fed on Wire Mother." The graphs track how much time that the baby monkeys spent with each mother in a typical twenty-four-hour period. What makes the charts so remarkable is how alike they are. By the age of six months, both groups are spending pretty much all their time, about eighteen hours a day, with the cloth mom. The wire-fed monkeys hustle back to the wire mother for food, but they eat fast. The charts show that they spend no more than an hour a day on wire mom. Mostly, every one of the baby monkeys are sleeping on cloth mom. Or cuddling. Or tucking their bodies close against her when they are startled. Or just stroking her. The graphs seem to have invisible writing running through them that says food is sustenance but a good hug is life itself.

Harry knew that he should summarize these results in the psychological jargon of the day. In that famous speech to the APA, he put it like this: "These data make it obvious that contact comfort is a variable of overwhelming importance in the development of affectional responses, whereas lactation is a variable of negligible importance." Psychology may have been insisting for decades that the baby's connection to its mother was a limpet-like grip on its source of food, but did psychologists really believe that? Could anyone watch the way an infant burrowed happily into the arms of a parent and believe that it was merely about milk? Or, again, in the lingo of the profession: "This is an inadequate mechanism to account for the persistence of infant-maternal ties." Really, do you believe that a lifetime bond is built on who holds the bottle? There had been no good experiments to justify that position, Harry argued. And now, when his laboratory had actually done the experiments, he and his students hadn't found a trace of a bottle-built relationship. They'd found that the baby monkeys—and Harry thought they represented human babies perfectly in this study—responded instead to the reassurance of a gentle touch.

Harry called this "contact comfort." He meant contact in its most nurturing sense, joyfully skin to skin. Comfort was just another word for security. "One function of the real mother, human or subhuman, and presumably of a mother surrogate, is to provide a haven of safety for the infant in times of fear and danger." When a child is frightened or sick, it instinctively seeks that haven, he said, and "this selective responsiveness in times of distress, disturbance or danger may be used as a measure of the strength" of the emotional connection. This again was an idea very close to John Bowlby's notion that a parent provides a secure base for a child. Harry and his students decided to pursue that connection further, to see whether they could better define the way a parent makes a child feel safe. What does it take to provide a safe harbor?

"We started asking simple questions," says Bob Zimmermann. "What does a baby do? How does a baby react in relation to his mother?" They were already beginning to appreciate just how intensely the small animals attached to the green-smiling surrogates. The surrogates were wrapped in a cloth "smock" that was changed every day for hygiene reasons. When the surrogates needed to be cleaned, a door came down between the baby monkeys and their mothers. The lab crew would then provide the clean clothes. Meanwhile, the little monkeys wailed in dismay, pressed against the door, paced the cage looking for her. When the door came up again, the babies would plaster themselves against cloth mom, grasping onto her smock like a lifeline.

"So when he's frightened, does he look to his mother?" Zimmermann asks. Does he hold on with that same desperate "save me from the world" intensity? By now, Harry's graduate students were fully engaged in the research. They bought a little toy bear that marched and banged a small metal drum and a toy dog that barked—"Arf, arf, arf," Zimmermann demonstrates. The point of these jerky, strange, noisy creatures was that they would trigger a fear response in the little monkeys. The researchers would put the toy in a box, roll it up next to a cage where a baby monkey was housed with a cloth mom, open the box, and start the bear banging the drum.

"These monkeys were only five days old, they could hardly walk, but some of them in sheer fear would just fly across the space to the mother," says Zimmermann, his hands making a graceful arc in the air, replicating the airborne lift of baby to mother. "By the time they were just a few weeks old, say thirty days, when they could really get their feet under them, everybody was running to the cloth mother," clinging with both hands, burrowing their faces into that warm fluffy body, closing their eyes. Pretty soon, the cloth mother was simply base camp. The little monkeys slept on her. A young animal might leave and explore a little, but he'd always hurry back. Even while venturing around the cage, he'd check over his shoulder to make sure that she was still there, watching over him.

The scientists now had twelve monkeys in the study. Zimmermann and Lorna Smith were testing the infants four to five times a day. The lab crew was putting together a picture based on Harry's favorite kind of statistics—small, tight, personal, real. "There were some people very critical of the surrogate work because of the statistics," Zimmermann says. Critics complained that the study lacked large cohorts of animals and the statistical depth that comes with a big population. "And my answer to them was—the study didn't need elaborate statistics. It just made sense."

The effects were so strong, the connection between baby and surrogate so visible, that the scientists began to wonder about other ways to test that bond and the security that seemed to come with it.

Bill Mason had taken a pocket of a room in the lab, six feet by six feet, and created a play space for the monkeys. He called it an "open field." Mason saw the field as a space where monkeys could be challenged beyond the WGTA's ordered trays. He wanted to use puzzles and locks and toys to study curiosity and he wanted to give the monkeys room to explore. Bob Zimmermann started thinking about the space itself. What if the open field could be made into a strange little world, a place somehow scary to a baby monkey?

Zimmermann had read a 1943 paper by a Gestalt psychologist, Jean Arsenian, who had conducted an experiment with children in a playroom. Arsenian had spent hours observing toddlers when their mothers were in the room, watching how they played or simply fell still when their mothers left. "And she talked about the mother having a field of influence," Zimmermann says, and that field was like a charmed circle in which the baby could safely and happily play. If a child felt secure, it was as if she carried that charm with her so that she could wander freely away more easily than a less secure toddler. The playroom was a powerful way of demonstrating the mother's field of influence, Zimmermann thought. So what if they took the open field and they used it like an Arsenian playroom—would cloth mom also have a zone of influence, generate a charmed circle of safety?

By the time Harry gave his presidential talk to the American Psychological Association in September 1958, he and his students had been testing four of their surrogate-raised mothers twice a week for two months in the open-field test. Zimmermann's question seemed a charmed prediction. In the open-field room, the little monkeys rushed to the cloth mother, clutched her, rubbed her, and cuddled. The first few times the monkeys were in the room, they never once let cloth mom go, just held tight and barely looked up. But after a while, as long as she was there, the little animals would look around the space. They would begin to get interested. The youngsters might climb a little, push a puzzle piece around, go back to mom, wander out for a while, return to mom, chew on a toy, return to mom. Without her, though—if the scientists decided to keep cloth mom out of the room—literally, the infants were lost. They would screech, crouch, rock, suck their hands. "Baby monkeys would rush to the center of the room where the mother was customarily placed, looking for her, and then run rapidly from object to object, screaming and crying all the while," Harry said.

Wire mom was almost as hopeless as no mom. The wire mother could sit in the open-field room all day, but she had nothing to give in terms of reassurance. She was just one more unnerving object. The baby monkeys who went into the field with a wire mother, even those who were accustomed to being fed by her, looked like abandoned children. They were terrified out there in the strange little world of the open field. Even with wire mom sitting squarely and obviously in the middle of the room, they would turn instead to the wall, huddling in its shelter.

As it turned out, the behavior of the little monkeys in Harry's lab fit almost exactly with some inspired studies done by a young supporter of John Bowlby's, an Ohio-born psychologist named Mary Salter Ainsworth. Ainsworth had earned her degree at the University of Toronto. She trained under a scientist there, William Blatz, who was trying to make everyone, anyone, including his students, listen to his "security" theory, which argued, in part, that a child derives security from being near his parents. That sense of safety, Blatz argued, enables the child to go out and explore the world.

Ainsworth, who always described herself as rooted in insecurity, liked the theory so much that she focused her Ph.D. dissertation around it. In 1950, serendipity brought her and Bowlby together. Her husband got a job in England. She went with him and began job-hunting. Bowlby, by that time, was advertising in the newspapers for help and Ainsworth answered his ad. She spent more than three years working with Bowlby, James Robertson, the researcher who had filmed hospitalized children, and the new and precarious attachment team before again following her husband, this time to Uganda, where he had taken a teaching position.

Determined yet to do something with her psychology training, Ainsworth began spending time in the tribal villages, at first just watching. For all her time with Bowlby, she still thought that the Freudians might be right, that a baby merely bonded to the person who fed him, reinforced, as it were, by being fed. But in the Ganda village, she watched children and their mothers and she changed her mind. The relationship was absolutely specific. "The mother picked up the baby, the baby would stop crying, but if somebody else tried to pick him up at that point, he would continue to cry," Ainsworth said. The villagers might even share feeding duties, rocking duties, but, in the end, neither milk nor bed created the same bond. A child always knew mother from other.

Babies smiled differently for their mothers; their faces lit in a way that they didn't for strangers. They cooed and coaxed differently. If a mother walked out of the room, her baby would crawl after her. If she came into a room, baby scuttled joyfully in her direction. One of the central points in Bowlby's theory is that attachment involves a precise relationship—and an interactive one. The mother responds to the baby and the baby does her best to keep the mother doing that. She wants her mother close. Therefore, the infant works to bind and attach her mother. In return for all that effort, the baby is rewarded with security. A smile is a wonderful example, then, of a baby's pulling his mother in just a little tighter. "Can we doubt that the more and better an infant smiles the better he is loved and cared for?" Bowlby liked to say.

In that Ugandan village, Ainsworth could see exactly the kind of behaviors that Harry and his students were producing in their experiments. The babies would make short excursions away from their mothers. Then they would stop and check, crawl back to touch, or just smile, making sure that she was still there for them. Ainsworth put it like this: "The mother seems to provide a secure base from which these excursions can be made without anxiety." Ainsworth, too, began to wonder about the nature of security. What behavior makes the good mother, the one who puts a child's world right? It was clear from Harry's work that cuddling and comfort were essential blocks in building a secure base; therefore, the wire mother could never be a source of security, no matter how often she provided food. It didn't matter that wire mom never rejected or walked away from her child. By the very fact of her metallic nature, she was unable to provide emotional support.

What if you translated the wire mother—the cool, businesslike but available parent—into human terms? Harry once told of meeting a woman who, after hearing about his research, marched up to him and diagnosed herself as a wire mother; she was uncomfortable holding her children, she said, and she disliked the clutch of their hands. "It could have been worse; she could have been a wire wife," he joked. But he told the story to emphasize that there are such mothers out there; wire mom wasn't just a lab creation; she represented a style of parenting. And perhaps because of psychology's fixation on feeding and conditioning, researchers hadn't realized how wrong that kind of cold and distant parenting might be.

What happens to the child who must navigate through life without a parent who is willing, or able, to provide security? If there's no way for a baby to bind a parent to him, heart to heart, then what provides him with a sense of safety while he explores the big, bad world out there? Can a small boy, a little girl, ever achieve the everyday courage of curiosity if no one loves the child enough to hug back?

You could make the argument that Harry's wire-mothered monkeys, afraid to explore, to touch, even to look around, were a perfect case study in insecurity—or even insecure attachment. A few years later, in the early 1960s, Ainsworth began testing for such responses in children. She had left Africa (and her husband) and was working at Johns Hopkins University. In Maryland, Ainsworth worked out a plan to monitor the way children attach to their parents, to see some of the consequences of a stable connection or a fragile tie. Her "strange situation" tests—not so different from the open-field experiments in concept—were rigorously designed and detailed in their measurements. Like Harry Harlow, like John Bowlby, Mary Ainsworth realized just how nailed-to-the-floor the studies had to be to make psychology pay attention.

The strange-situation test is still used today and it works like this: Mother and infant arrive at the lab together and settle themselves in a playroom. A friendly researcher welcomes them and then sits quietly in a corner. Toys and games litter the cheerful room and, typically, once there, surrounded by all the bright plastic possibilities, the baby crawls off to explore. But here's the catch. A few minutes later, the mother leaves the room; the baby is now alone in that fascinating but still strange place. Only the unknown researcher remains. Then, after a few minutes or so, the mother returns.

Remember the way the baby monkeys leapt and clung to their cloth mothers? Almost all the human babies, too, rushed to the returning mother. They smiled and they clutched her close. If the mother's leaving had been a little impatient or brusque, the baby might even cling tighter on her return. Uncertain in their mother's absence, those children seemed to be testing her response to them. They wanted extra reassurance: They were glad to see her; was she happy to see them?

Without their mother, many of the children stopped playing. Some cried. They might tearfully look around, search the room, toddle toward the door in search of the missing parent. None of them found the researcher's presence reassuring. They tended to look at her doubtfully, if at all. She was a stranger. Nothing about her presence made them feel secure.

There were some variations on this pattern. Sometimes, a child would continue playing in his mother's absence and, upon her return, still relaxed, merely look up, beaming. Ainsworth classed these children as beautifully, securely attached. They seemed to have no worries that their mothers wouldn't return; they were just "there" for them. Ainsworth also found the opposite, responses that seemed to suggest the child of a wire mother. Some children showed no comfort or happiness when their mothers returned. They might crawl to the mother, as if seeking reassurance, but then hold their bodies stiffly away from her. Others didn't even try; there was an odd wariness in the relationship. They would flick the mother a glance and then look away. Here was born the term "insecure attachment." After further study, Ainsworth divided the insecurely attached children into two primary groups. There was the ambivalent attachment, such as the child who sought the hug and then couldn't really get anything from it. And there was the avoidant attachment, the slightly hostile connection between the mother and the child who knows her and looks away from her.

One of the enormous differences between these children and Harry's surrogate-mothered monkeys was the relationship itself, the give and take, back and forth between baby and mother. When Ainsworth and her students visited these families in their homes, again as observers, they found that the mothers of securely attached children were acutely tuned to their children. They were responsive to the cry and the smile, quicker to pick up a crying child, inclined to hold a baby longer and with more apparent pleasure. The mothers of ambivalent children were often unpredictable—some moments hurrying to cuddle, some moments indifferent to a baby's sobs. Neither the researchers nor, apparently, the infants could rely on which response a mother would give. The mothers of avoidant children might be called rejecting in manner. They were often irritated when they did pick up a child. They did it resentfully and a little roughly. They sometimes spoke of their dislike of physical contact, and in Ainsworth's tapes they could be heard snapping "don't touch me" if their children reached out.

Ainsworth's findings countered Watson almost point by point: mothers who responded quickly and warmly to their babies' cries during the early months of life not only had babies who were securely attached but also babies who actually cried less than the others. By age one, those youngsters seemed to feel they didn't need to cry for attention. They relied on gestures and expressions and gurgling coos. They didn't look clingy, as the old theory would have predicted. They looked independent. By contrast, the avoidant babies apparently had learned to expect nothing from their parents. The infants sobbed frequently and usually to themselves. They didn't look self-reliant. They looked miserable—or angry. And if their mothers did come to pick them up, the defensive infants turned away. Huddled away like that, the avoidently attached babies resembled nothing so much as Harry's wire-mothered monkeys, lost in the open-field room, heads down against the wall.

Could Harry's work—or the timing of it—have been more perfect for John Bowlby's arguments? Alienated from psychiatry and the high priests of human behavior, Bowlby had already been spending more and more time talking to animal behavior researchers. Among the best was Konrad Lorenz, who would later win the Nobel Prize for his work with "imprinting"—the passionate, instinctive attachment of baby birds to their mothers. Lorenz was able to show that this first loyalty was given to the "mother" first seen by the tiny birds. The "mother" thus could be Lorenz, actually, if he was hovering over the nest when greylag goslings—the species he studied—first cracked their way out of the egg shells. It wasn't a perfect match, obviously, for human behavior. It certainly didn't impress Bowlby's critics. "What's the use to analyze a goose?" mocked one of Bowlby's colleagues in the British Psychiatric Association.

If you considered Lorenz's studies seriously, though, you realized they were a reminder that nature fully intends a helpless baby to be well connected to a protector. In the geese, the attachment might seem hard-wired. Human relationships are more flexible, and therefore more difficult, but Bowlby insisted that the basic point was the same: Mothers mattered. Babies needed them; babies were born to need. And now here was Harry Harlow, conducting experiments on a species much closer to humans. Harry's work was saying exactly the same thing. The Wisconsin experiments blew away the notion that mothering was equivalent to feeding, or that any old mother could comfort a child. You might not like Harry's results, but you couldn't ignore them.

John Bowlby wrote to Harry Harlow on August 8, 1957, after the cloth and wire mother work had started but before Harry drafted his landmark speech. Bowlby had heard of the Wisconsin surrogate experiments from a highly respected animal behaviorist at Cambridge University named Robert Hinde. At a psychology meeting at Stanford that year, Harlow and Hinde had fallen into an extended discussion of motherhood. The discussion made enough of an impression on Hinde that, upon his return to England, he promptly contacted Bowlby. And what Hinde said intrigued Bowlby enough that he promptly sent Harry a draft of a paper he was working on: "I need hardly say I would be most grateful for any comments and criticisms you cared to make . . . Robert Hinde told me of your experimental work on maternal responses in monkeys. If you have any papers or typescripts, I would be very grateful for them."

The draft manuscript Bowlby enclosed was that near-diatribe on behalf of children, the report that would infuriate so many of his psychiatry colleagues in London, *The Nature of the Child's Tie to His Mother.* Unlike the British psychiatrists, Harry loved what Bowlby had to say. He wrote back promptly himself: "It appears that your interests are closely akin to a research program I am developing on maternal responses in monkeys." He invited Bowlby to come and see the surrogate-raised monkeys. Harry wrote that the relationships they were looking at were growing stronger and stronger, beyond anything that researchers in his lab had expected.

In one series of tests, Harry's graduate students had put a cloth mother into a Plexiglas box in the middle of the open-field room. The baby monkey clearly didn't want mother behind glass; she cooed coaxingly, she fingered her way around the clear barrier, trying to find a way to get her parent out. But she would, in the end, settle for just being able to see mother's face. All the little monkeys tested would eventually begin exploring the room, and when they found something that interested them—a small puzzle, say—they would pick it up and bring it back to the box, as close to mom as they could get. In other tests, the monkeys could open the box if they figured out how to undo a series of locks. They would puzzle and puzzle without end until they had every lock undone and could tuck themselves close against mother. Even if young monkeys had graduated from cloth mom's care and been moved into the company of other juvenile macaques, they would leap to free her. They might have been away from her for months, but it made no difference. Apparently, Harry said, the little monkeys were "resistant to forgetting."

Bowlby promptly began citing Harry's work. He would say later that the two research projects that began to make people take him seriously, that eventually eased him back into the British Psychiatric Association, were the stunning work of Mary Ainsworth and the inarguable findings from Harry Harlow's lab. After them, Bowlby said, "Nothing more was heard of the inherent implausibility of our hypotheses."

The speech Harry made in 1958 upon assuming presidency of the American Psychological Association rings like a war cry: exasperated, provocative, and startlingly poetic in its very outrage. He wondered out loud that his profession could be so willfully blind. "Psychologists, at least psychologists who write textbooks, not only show no interest in the origin and development of love and affection, but they seem to be unaware of its existence." The only books that seemed to address love were written by fiction writers, by poets and novelists, and they were fixated on adult love. It was as if the whole world were colluding to pretend that our first loves, those of childhood, don't matter at all.

And when Harry Harlow described the cloth mother experiments he touched on Bowlby's eloquent explanations of why the child clings to the mother. Everything, however preliminary and new, and contrary to earlier arguments, spoke of a "deep and abiding bond between mother and child." The poets and the novelists, he said, might have written more prolifically and more beautifully of love but, in the end, he thought they could not always illuminate it as well as a scientist with a willing mind. "These authors and authorities have stolen love from the child and infant and made it the exclusive property of the adolescent and adult," he said. In that, he could promise his audience, the poets were as wrong as the psychologists. Love begins at the beginning; perhaps no one does it better, or needs it more, than a child.

SEVEN

Chains of Love

Because of the dearth of experimentation, theories about the fundamental nature of affection have evolved at the level of observation, intuition and discerning guesswork.

Harry F. Harlow,
***The Nature of Love,* 1958**

IN THE LATE 1950S, A trio of child psychologists—Joseph Stone, Henrietta Smith, and Lois Murphy—decided to pull together a book on the science of babies. They started collecting studies. Then more studies. And still more studies. The paper pile was getting so big that, Stone said, they started wondering whether they were merely incompetent researchers, unable to get a grip on the science. Finally, they realized that the stacks of studies held a unified message: They had tapped into a "genuine knowledge explosion." Science was finally discarding its vision of the passive child. Suddenly, babies were real people, people with feelings, and passionate ones at that.

The resulting book, *The Competent Infant,* began with a story: "Some years ago, a young psychologist of our acquaintance was helpfully diapering his six-month-old first born son. His wife came on the scene and protested: 'You don't have to be so grim about it, you can talk to him and smile a little.' At which our friend drew himself up and said firmly, 'He has nothing to say to me and I have nothing to say to him.'"

And how wrong he was, the authors said. And how wrong science had been. They followed that assertion with a two-page list of other mistaken assumptions adopted by baby experts about babies. As they put it, "We can collect embarrassing moments from the professional literature almost at random." Among their choices for the most ridiculous scientific ideas: Babies can't see faces (reported in 1942); babies are unaware of almost anything around them (1948); newborn babies are only a collection of reflexes (1952); they don't see color until the age of three (1964); and even "the human infant at birth and for a varying period time afterwards [is] functionally decorticated" (1964); in other words, babies are brainless.

The Competent Infant is a 1,314-page, fully loaded rebuttal to the empty-headed infant idea. Pointed commentary from the editors is wrapped around 202 studies and essays authored by scientists from around the world; each contributor aimed at reinventing our image of the child. The babies in Stone and Smith and Murphy's sharply edited book can see people just fine. They pay attention to those people, too, and think about them. These small humans work hard at relationships. Parents matter; and, oh, by the way, love matters, too.

Stone, who was chair of the child study division at Vassar, and his colleagues picked their evidence carefully. They realized, as Bowlby had before them, that to make their argument they were going to need not only direct human evidence but also circumstantial evidence, the kind found in well-controlled animal studies. Still, the research had to be the best; they included only a tenth of the studies they had gathered. Stone contacted Harry Harlow almost as soon as *The Nature of Love* came sizzling off the APA presses. The cloth-mother studies—and Harry's outspoken championship of love and relationship—had catapulted Harry out of the small community of primate researchers into the bright-light, big-city world of baby care expertise.

It wasn't only Stone, although Harry liked Stone; he called him "one of nature's wonderful men" and provided him with reams of material for the baby book. At that moment, everyone seemed to be calling Harry F. Harlow to hear his message about love. He spoke at campuses around the country, hurrying from conference to committee meeting to conference. And Harry was also talking mother love and baby love outside the scientific community: he appeared on the television networks, in national magazines, in newspapers, on the talk circuit. *Life,* the *New York Times,* the *Los Angeles Times,* the *Washington Post,* the Associated Press, CBS, NBC, the BBC—they all wanted him to reinvent the mother-child bond. One of the reasons, his former colleagues agree, why Harry Harlow was so effective in shifting mainstream psychology's stand on love was that he was such a tireless and eloquent advocate for his cause.

"Harry had already established a reputation as being bold, caustically witty, playful and innovative—qualities that were very rare among academic psychologists," says Bill Mason. "He had also demonstrated that he was eager to apply these qualities to attacking some of the sacred cows of the Hullians and Skinnerians, central figures in the Zeitgeist of experimental psychology of the fifties." Mason believes that the message was well timed. There was a related uprising of work by ethologists such as Konrad Lorenz, Bowlby's colleague Robert Hinde, and University of California psychologist Frank Beach, all illuminating the importance of early experience. Donald Hebb, too, had performed rat studies showing that early experience could alter adult performance in animals. Their studies and Harry's spoke to many who had become uneasy about the rigid direction psychology seemed to be taking. It was an unusually receptive moment, a rare opportunity to reconsider the mainstream views of psychology.

Skeptics couldn't argue away the detailed, beautifully graphed primate studies from the Wisconsin lab as they had done with the more anecdotal human evidence cited by Bowlby and others before him. "With the human data you could always argue that there were unknown early experiences that had shaped them or that the populations had been preselected by the very factors which the scientists were finding. For example, if you find that kids in a hospital ward are sicker than kids at home with their families, it does not mean that being in a hospital ward produces sickness," explains University of Georgia psychologist Irwin Bernstein, himself a long-time expert in the social behavior of monkeys. Critics couldn't so easily dismiss the controlled, confirmed, and reconfirmed results from the surrogate mother studies. Wire mom's inability to nurture, well, anything, just couldn't be argued away.

"Of course, Bowlby could hardly deliberately put a human baby on a wire surrogate," says Robert Zimmermann, one of the first collaborators in the Wisconsin cloth-mother studies. Beyond that, Zimmermann and Mason both agree that Harry's belief that science was about life made him unusually effective in getting his points across. Harry would talk to anyone. He fiercely resisted operating on an upper-level, theoretical plane. "Bowlby was a white-collar scientist," Zimmermann says. "Harry was a blue-collar scientist." Mason's analysis echoes that: "Bowlby's style was more scholarly, more technical, more difficult to read, and more tentative than the *Nature of Love,* and it did not offer a simple take-home message. It was not picked up by the popular press nor as widely circulated among professionals." Bowlby wasn't a natural salesman. A talent for persuasiveness is not necessarily admired or cultivated by scientists. By this reckoning, Bowlby's dry and dignified presentation helped suppress the attention given to his work. It was the showman coming up behind Bowlby in the wings—Harry Harlow with his user-friendly approach to science—who shifted the spotlight back toward the theory of attachment.

Indeed, at that moment, much of Harry's energy was given to illuminating the message, making people listen, talking to scientists, appearing on CBS's *Conquest,* telling the world that love mattered. Some of his colleagues accused him of grandstanding. Being Harry, he never allowed himself to get too carried away in his role as messenger. He was amused to find that psychology seemed to be catching up with the basic common sense of many parents. Mothers and fathers, those who had never paid attention to scientific opinions, were already aware that good mothers hug their children and that good fathers take time to play with their sons and daughters. Bob Zimmermann recalls Harlow's return after he first took his cloth-mother research on the road. Harry was grinning, shaking his head. "Mothers know, Bob," he said, amused that science was once again chasing basic common sense. In an interview with the *St. Louis Post-Dispatch,* Harry joked that his lab would be designing a new, better surrogate. It would have ten arms, which would give it the minimum number of hands needed to provide adequate child-care. "It was a mother," Harry said, "who suggested the number."

Professor Harlow was suddenly consulted on a startling range of child-rearing techniques; not just love, but everything from naps to toilet training. Ever opinionated, he expressed his doubts over the then widely held belief that children should be toilet trained before their first birthdays. His response was classic Harry, common sense balanced with humor: "There are many techniques for early toilet training of human infants, and all have one thing in common—they do not work." Let the child grow up and mature a little, take some time to enjoy having a baby, Harry recommended. "Masochistic mothers are, of course, welcome to enter the ordeal as early as they wish."

He was more delighted when his work was cited in a campaign by *Redbook* magazine to persuade hospitals to let newborn babies and their mothers stay together. The editors named Harry's work as proof that hospital procedures were good for neither parent nor child. *Redbook* was blunt: "When mother's needs and hospital regulations collide head on, there can only be one acceptable answer—the regulations and customs must give way." Another glossy publication, *This Week* (a now defunct competitor with the Sunday newspaper insert magazine, *Parade*) suggested that thanks to Harlow's work, American doctors now needed education in primatology.

Harry told the magazine that monkeys did indeed offer insights on child rearing. Perhaps most important, he said, is that "learning to love, like learning to walk or talk, can't be put off too long without crippling effects."

Stone invited Harry to Vassar in 1959. That same year, Nancy Bayley, the famed UC-Berkeley child psychologist, wrote to Harry to say that she had nominated him as a fellow in the APA's division on developmental psychology. He wrote back to her with pleasure: "To say that I have gone developmental is almost an understatement."

Harry was enthusiastic about *The Competent Infant* from the beginning. He wanted people to see that primate work could support notions of human behavior. And he wanted psychology to change. His surrogate work was a starting point, an early demonstration, as Stone put it, that the "infant does not live for bread (or milk) alone." Elegantly cited as that was, if it didn't signal a change in psychology, then the cloth-mother studies would be a failure. The real measure of Harry's success would not just be in the sparkling acclaim of his time. To succeed, the field would have to go beyond momentary interest into something more substantial. There would have to be more than a flurry of stories in the popular press; scientists and public alike had to be willing to consider that relationships mattered from the first day of life. And they had to consider that babies knew this. In mother-child relationships, perhaps psychologists even needed to confront the idea that babies knew more than they did.

Stone and his coauthors did see the early glimmer of change toward the Harry Harlow view, at least when it came to the scientific perspective on how a baby connects with—sees, touches, needs—his mother. The editors of *The Competent Infant* thought there might even be a new consensus that "babies do not love their mothers because their mothers feed them; [that] it might be more accurate to say that infants become addicted to their mothers." And in this, indeed, was a hint of a revolution in the making, an increasing acknowledgment of the power of a single relationship.

It's a potent idea, that fleeting moment when we are, briefly, an addictive substance—totally loved, totally needed. We may achieve such intensity in first heat of a love affair. But it may matter more that we accomplish that be-all connection in the dawning days of life. Mothers, wrote Stone and his colleagues, are the first mesmerizing objects in a baby's world. And even after the infant has learned to recognize and value others, a baby will still single out his mother's face. She often remains the best-loved face for a very good reason. She's the most *there,* the most responsive of any person in the baby's world. As psychology moved away from the notion of a baby as a passive recipient—bonded blindly to the bottle or the breast—the possibility of interaction between parent and child drew sudden attention.

Harry's cloth-mother-raised monkeys gave the idea unexpected emphasis. In the first days, the cuddly smiling cloth mother had looked as if it might be science's answer to parenting. But consider the plight of a baby who is addicted to a statue. As scientists at Wisconsin and elsewhere watched the surrogate-mothered monkeys grow up, they became increasingly aware that the animals were in trouble. One New York reporter, visiting Madison for a follow-up story on surrogate mothers, described the one-hundred-odd monkeys with real dismay. Some clung to the bars of their cages and shrieked at passersby. Others mauled themselves, biting their arms, ripping out fur. "Many more present an unnerving picture of patient apathy," the reporter observed. "Hour after hour, they sit in strangely contorted positions or huddle in the corners of their cages, seeming to see nothing, seeming to hear nothing."

The worst of these—the most desperate of the rockers and biters—were those who had been raised by a wire mom. But the children of cloth moms were also unexpectedly dysfunctional animals. Cloth mom had seemed so warm, so cozy, and so cheerfully available. At first, the researchers weren't even sure what had gone wrong. It seems obvious in retrospect; it was less so in that uncertain time, when science was barely beginning to think about the connection between individuals and their relationships. The children of cloth mother forced Harry and his team of students to reconsider the two-way nature of a bond between two beings. Once they had, the scientists could clearly see the nature of those lost monkeys. They were babies who hugged their mothers and were never hugged back.

It doesn't take a scientist to know that there's no future in a relationship that is only addictive for one of the two people in it. Let's propose that all our relationships, be they in infancy or old age or anywhere in between, work best when we actually pay attention to each other. The question that might follow is whether any theory of behavior can stretch to encompass the ever-flexible, complex, and shifting quality of a relationship? Or does it need a theory? Is it attachment behavior or is it just good healthy interaction when we watch a partner's expressions, listen carefully, laugh once again when our children tell the same joke for the eleventh time, notice a bare flicker of sadness and move to comfort? Perhaps the two—attachment and interaction—can't be sliced apart quite so neatly. Or at all. Bowlby always argued that part of building a secure attachment was the way the mother responded to her child—and vice versa. But there is also a distinct line of research that explores give-and-take without worrying about a full-blown theory behind it. In the first arguments of scientists over what defines a relationship, one can again see the fine breath of change just stirring the edges of psychology.

Some of the most intriguing science of, let's say, responsiveness began with the ever useful rat. Those studies, conducted by innovative behavioral researchers such as Seymour "Gig" Levine, grew, as Harry's mother love studies had, out of the 1950s. Levine's work also supports the smart baby presented in *The Competent Infant.* Levine, who remains an advocate for the two-sided nature of all relationships, keeps the mother-child bond as a part of the story but not the whole story. "I've never really liked attachment theory," admits Levine, a professor of psychiatry at the University of California-Davis. "And that's not meant to say that the mother-infant relationship isn't important, especially in species like our own."

When he first started thinking about the power of relationships, Levine was a big, dark-haired New Yorker with clear hazel eyes, a rapid way of sifting through ideas, and almost no tolerance for tedium. He started tinkering with the mother-infant rat relationship after graduating with a psychology Ph.D. from New York University in 1952. Levine had begun by doing sight and hearing research, but that had bored him mindless. He'd pretty much decided to junk the whole profession of experimental psychology and become a therapist. With that thought, Levine grouchily took a research position in the psychiatry unit at Michael Reese Hospital in Chicago, where he found himself in the company of some of the best endocrinologists in the country. They were exploring the then-startling idea that experience could alter body chemistry. In particular, the Michael Reese scientists were trying to follow the bright, shifting paths of hormones in the nervous system. It was such an intriguing idea that "the whole world opened up for me," Levine says. The senior scientists at Michael Reese were interested in the body's stress response. People knew, of course, that unhappiness and stress impaired disease-fighting abilities; hadn't they seen that very thing in the hospitalized children? But they had no clear idea how that happened and why.

Levine decided to tackle stress in an animal model, beginning with the rat. He was curious about whether events during infancy led to the development of psychological disorders. He came up with a very simple experiment—it would compare three groups of infant rats. The first—and theoretically the luckiest—would simply stay cuddled into their cage with mother. The second would be taken out once and exposed for three minutes to a mild electric shock, "as a model of early trauma." The third group would also be removed for three minutes, but only to another location. For comparison purposes, they would also be "handled" but not shocked. Levine wasn't thinking about relationships particularly; he was just looking for a good stress response. He applied for a federal grant to support his experiment. The scientist who came to review his proposal was a slight, sardonic Wisconsin psychologist, Harry Harlow, none other, who thought the work sounded pretty damn interesting and successfully recommended $8,000 in funding. "He was very supportive," Levine says. "This was about the same time he was doing the monkey work, so he got the experiment immediately."

Levine had a straightforward idea. He would create a model of a sheltered infancy compared to a stressed-out one. Then he would observe both sets of animals as adults. Would the traumatized infants grow into neurotics? Would the cozy homebound rats mature into extra confident rodents? As the rats aged, he ran all three groups through a simple stress test, an electric shock that could be ended by simply skittering away from the charged floor grid. The results contradicted every expectation. The non-handled rats looked almost the opposite of confidant. They were the slowest learners. They were the most insecure. The tingle of the electric shock agitated them; they could barely bumble their way to safe ground. By contrast, the handled and even the shocked rats grew into pretty cool customers. They were quick to respond, sure-footed in their move to escape the shock, and they displayed far less angst. It wasn't only the reversed effect in behavior that startled the scientists. "We all looked at each other and said, 'How the hell could three minutes have such a profound effect?'" Levine recalls.

The potency of a 180-second time-out served as a reminder that a living creature must be exquisitely tuned to its environment. Apparently, the body generates a reaction even to the slightest change in circumstance. The scientists began to suspect that the early handling of the rats, the early exposure to stress, altered the neural circuitry of the brain. The experience itself might be brief, but the effect—generated by some unknown tweaking of brain chemistry—was surprisingly long lasting.

What was going on? Levine wasn't sure. He decided to explore further, this time with another version of the strange-situation test. For the rats in his research project, this meant being sent into an open-field test. To a rodent, a wide-open area can be perilous terrain, offering no possibility of shelter to a small animal always poised against the possibility of predators. Levine's tests alternated that dangerously empty space with an open field spiked with new objects and toys, much like the open field used in Harry's lab. Levine had a growing list of questions. Would the rats panic at being so vulnerable, perhaps freeze in place? If he placed some interesting toys in the room, would they be curious enough to explore?

Again it was the mom-sheltered rats that fumbled the challenges. They froze. They peed uncontrollably. They trembled at the sight of a new object. Later experiments showed that even when scientists put a treat in the open field, such as sugar water, the sheltered rodents were afraid to check it out. It was the handled rats who sucked up the treat. The non-handled rats never seemed quite sure that they could trust their senses. It appeared they were overwhelmed by their agitation when anything new or different happened, so much so that they could hardly make a reasonable choice. They seemed perpetually tuned to the anxiety channel in their brains. The early-stressed rats, by contrast, seemed to be listening in on the easy rock station. They looked unfazed by new situations—and they looked smarter. This made absolutely no sense, at least not until the scientists had rethought some of their ideas about what goes into a healthy childhood.

Levine and his colleagues came to appreciate that babies—even baby rats—need interest and interaction in their lives. Even a stressful interaction can be better than no interaction at all. Stress, in this case, does not mean an extreme situation, such as an abusive parent. It refers to everyday tensions. The truth is that even the most loving parents succeed in stressing their babies. Infants are picked up when they are sleepy and passed about so that strangers can admire them. They are fed strained spinach when they really want apricots. They are plunged into baths when they were happy being dirty, thank you. Interesting objects are taken out of their mouths. Baby life is, when you think about it, a daily barrage of frustrations. And maybe, too, if you think about it, all these little hassles prepare you for the big hassles, toughen you up gradually, are comparable to putting your stress system through a few healthy push-ups and bench presses.

When the infant rats were moved around, their stress hormones went up. The brain was forced to respond to that, to regulate the response, to fine tune the stress level itself. The workout seemed to make the brain more flexible, so that the baby rats who were handled were better able to adjust to later demands. The more sheltered rats seemed to come up with an all-or-nothing stress response. They were almost equally upset by the maze and the electric sizzle, which turns out to be not a particularly helpful way of reacting. And to Levine, this strongly suggested that some challenging experience—as long as it's not an overwhelming traumatic experience—may be helpful, even necessary, in building the kind of body response system that will serve us best.

Does this mean that being mothered is bad for you? Does it call up that old stereotype of the overprotective mother? Not at all. If you compare the two groups of rats carefully, you realize that the stay-at-homers weren't living in some well-mothered paradise. They were spending their infancy locked in a box with a bored and often apathetic adult female. There's nothing normal about that. In nature, the mother moves around, leaves to hunt for food. Other rats come and go, and the odd predator makes a terrifying appearance. In a small cage, food and water placed right there, the world becomes pretty small and dull. The mother can become lethargic. Remember that the handled rats weren't undergoing lengthy separations from their mothers, either; they were spending three minutes out of every twenty-four hours away from home.

The researchers began to suspect that the subdued response of the stay-at-home rats grew from what they called "restricted experience in infancy." This meant the babies had nothing much to do, nowhere to go, and too often a mother boxed and bored into indifference. They were emotionally unprepared for the rest of the world. By contrast, the early handled rats were less emotional, perhaps due to a "toughening up process as a result of experiencing stress during development." Later, Levine and his colleagues would explore further the reasons for that difference in emotional response. The brief separation from their infants, that potent three minutes, sometimes woke the mothers up into a kind of renewed affection. There was a sudden intensity to their interest in the rat pups when their infants returned. The babies were licked and sniffed and rubbed and, well, nurtured. It appeared that a brief separation—if it did, indeed, make the heart grow fonder—could actually improve a relationship.

None of us—and we may thank the fates for this—evolved in a world as static as that of a little rat in a lab cage. Levine suspects that our bodies, right down to our hormonal systems, are beautifully designed for coping with challenge. But not so well designed for bland nothingness. We—whether human or rat—evolved and adapted for the messiness of reality. Levine is eloquent on the point:

> Whenever a parent picks up a baby, or a child tussles with his puppy, there is some stress involved. Almost all experiences of infancy involve some handling by a parent or some other larger and supremely powerful figure. Even the tenderest handling must at times be the occasion of emotional stress. Perhaps the only children insulated from such experience are those reared in orphanages and other institutions, and the only animals, those that live in laboratories. In the ordinary world, the infant must grow under the changing pressures and sudden challenges of an inconsistent environment.

Within a few years, at least among rat researchers in the Midwest in the mid-twentieth century, this was a hot idea indeed. At Indiana's Purdue University, psychologist Victor Denenberg's studies emphasized exactly what Levine was finding, that adding a little spice into the life of a young rat made a big difference in the adult animal, that "infantile handling resulted in rats which were more exploratory, or curious, than those left alone." Further, Denenberg would go on to to show that this could be a generational effect. Early-handled female rats seemed to be much more engaged parents. Their pups, allowed to stay with their nurturing mothers, looked as adept and cool-ended as those handled rats from Levine's first study.

Levine began sharing his work at scientific meetings and promptly had a Bowlby-Spitz-Robertson-Harlow kind of experience. He was assured that he was wrong. When he presented his findings at the 1961 International Congress of Endocrinology, he argued that experience in infancy was a potent influence on adult behavior. "People didn't believe it," he says. "They were absolutely incredulous." Levine is philosophical about this: "It was an odd mentality but it was just the way things were. I didn't think of it as particularly strange that no one thought mother-infant interaction was important. That was the world I knew. I just kept on doing it. I knew my data and they were reliable, they were replicable and they were consistent."

One of the reasons, of course, that Levine sees his work as separate from that of Bowlby and his followers is that rats are a perfectly awful model for attachment theory. There's no beloved face of the one important caretaker in the rat world. Rat females mother when their hormones tell them to. Turn off the hormones and the flow of mothering shuts down, too. Mother rats aren't so fussy about who's sleeping in the nursery. When Denenberg performed an experiment in which rat mothers took care of baby mice, there were no objections or concerns from either rodent. The interesting part of that study, actually, was that being raised by rats—smarter, more social, and less snappy rodents—created unusually nice mice.

"You just can't apply attachment to rodents," Levine says, because baby rodents don't pine for a specific mother, nor do mother rats bond to their offspring as if they were the only babies in the world. As Harry had shown in his dissertation work, mother rats are intensely watchful and engaged caretakers. When separated, they'll doggedly round up the infants for care and feeding. But you can add new baby rats or remove them without really troubling the family structure. Harry's early look at rat parenting, touched on in his dissertation, testified to the power of maternal instinct, but *not* to the bond between one mother and one specific baby.

One of the most important lessons of mother-infant relationships, at least from the perspective of animal research, is that you must take them species by species. Rats connect one way, monkeys another. But even that is too simple a description. Different monkeys hardly mirror each other, either. Levine later worked with squirrel monkeys—elegant gold-and-gray South American aerialists—and found that the one-to-one relationship, mother to her own child, was so strong that if he tape-recorded one infant's wail of distress and replayed it to a group of monkeys, he could instantly pick out that baby's mom: She was the one leaping up in dismay. Bill Mason went on to study another South American species, the fluffy titi monkey, in which adult males and females form a lifelong partnership. For those monkeys, the adult partner relationship appears far more important than their attachment to their offspring. Mason found that stress hormones rose far higher if a titi mother was separated from her male partner than from her baby. There's often a readable message in body chemistry. This study showed the titi mother's hormone as a bright flare highlighting the most important relationship—and it wasn't the mother-child bond in the slightest.

Another of Harry's former graduate students, Steve Suomi, tested "secure base" responses in capuchin monkeys. Suomi gained his Ph.D., and a life-long interest in complex relationships, while at Wisconsin. He now directs a primate behavioral research program for the National Institutes of Health. He was startled by the capuchin study, "surprised to realize that when you take this very smart and successful South American monkey, and look at its relationships, you don't see the secure-base phenomenon." Baby capuchins don't draw the same sense of security from their mothers as other monkeys. "I've wondered what Bowlby's attachment theory would have looked like if Harry had been studying capuchins at that time," Suomi says. Because in rhesus macaques—as in, apparently, the human species—the warm and giving and reactive mother is exactly the one who follows nature's design.

Harry began to consider the tantalizing, baffling shape of mother nature an almost personal challenge. Was it possible, he wondered, to create an inanimate mother who could meet all of a baby's needs? Maybe, maybe not. But by pursuing that goal, psychologists might be led to a better scientific measurement of motherhood itself. The Wisconsin lab crew decided to take that challenge, start at the beginning, and work through the minimum requirements of motherhood. They focused on two material aspects in particular: warmth and motion. Does body heat make a difference in mothering? Is motion—being rocked and carried—essential to baby care?

"We created some hot mamma surrogates," Harry explained in one scientific report. To do this, the researchers replaced the usual faint warmth from the light bulb placed behind the surrogate's body with heating coils enclosed inside the body. This created surrogate mothers that were about 10 degrees Fahrenheit hotter than the usual ones. The increased temperature sort of made a difference. The babies preferred the hot body, but not in a winner-take-all, you-or-nothing kind of way. The little animals would still cuddle happily enough with the lukewarm mother. So did the researchers have it wrong? Did warmth matter or not?

Perhaps they were asking the right questions in the wrong way. Suomi decided to compare the hot mamma with a cold mamma. In this cooled-down model, the terrycloth sheath covered a hollow shell with chilled water circulating in it. These were not extremes; hot mamma was 7 degrees Fahrenheit above room temperature, cold mamma about 5 degrees below. But the baby monkeys reacted to the cold mamma as if she had been carved from some Arctic glacier. One little female, first placed with the warm mother, reacted to the cold substitute by leaping into a corner and screeching. Suomi then put a little male in with the cold mother, and that baby monkey responded by adopting a position of total rejection. He turned away from all surrogate mothers on sight; wouldn't even try out a hot mamma when they put her in the cage.

Does this mean that babies simply don't like an icy touch? Who does? Suomi thinks that the response is more than just preference, though. The love of warmth is probably like the need to cling, Suomi says; it's part of staying alive. Temperature is vital to survival; very small monkeys—and very young human babies—can't keep themselves warm enough, can't regulate body temperature. The lightbulb-warmed surrogate provided adequate heat, so that there was no dramatic reason to prefer the hot mamma. But cold mamma could be physically dangerous. Rejection and even despair thus seem a valid response to the chilly parent.

In the mechanics of being a good mother, then, it seems that keeping baby warm matters. Is it equally helpful to be in motion, a moving mother? You couldn't remotely describe those cloth surrogates as anything but statues. The living mother's chest swells with her breath, thumps with the beat of her heart, shimmers with the flex of muscles under her skin. Her arms cuddle and rock and carry. If she's a monkey mother, the baby clings to her fur as she scrambles up trees, over rocks, across feathery grasslands. We humans hug our children tightly, toss and tickle them; and even when a mother stands in one place, there seems to be an irresistible urge to sway, to rock gently back and forth. Is there a hint in that almost compulsive sway that our biology dislikes a motionless mother?

As with hot mammas, the studies in Harry's lab first showed only that small monkeys had a preference. The baby monkeys liked a rocking surrogate more than a stationary one; they ran to it more often and clung longer. Later on, the Wisconsin researchers experimented with a "swinging mother" that dangled about two inches above the floor of the cage. The swinging mother seemed to be even more loved, more tightly clasped, by the little monkeys. There was something about her wobbly presence that seemed to make them feel safer. Swinging mom made a better secure base; monkeys raised with her were braver in the open field tests than the children of an ordinary cloth mom.

It took a while for anyone from Harry's lab to follow the moving mother trail. But after he left Wisconsin, during a stint at primate facilities in Florida and Louisiana, Bill Mason returned to it. By that time, Mason had young scientists working under him. One of them,

Gershon Berkson, joined him in taking a second, harder look at mothers who move. In a study published in the mid-1970s, Mason and Berkson compared baby monkeys with a stationary surrogate to those raised with one that rocked and turned and jiggled the baby around.

Their results were provocative. It wasn't just that babies were happier with a mother in motion. They were better adjusted physically. Those raised with the mother that moved didn't exhibit the physical peculiarities so often seen in surrogate-mothered monkeys. They didn't rock and clasp themselves and huddle against the cage walls. They acted like monkeys from a good home. But why? Clearly this mother didn't hug back, either. So what was it about those bobbles and jumps that kept the little monkeys in some kind of balance? Berkson, now a psychology professor at the University of Illinois in Champaign, has continued to pursue the healing power of motion. He suspects, like other researchers now, that part of the answer does indeed have to do with the simple mechanics of mothering.

It seems that the swish and swing of parent rocking baby, parent carrying baby, is needed to induce normal development. It's comparable to the way a little healthy stress pushes the brain to grow appropriately. Motion nudges the nervous system. It's forced, again, to respond to the body's sudden instability. The nerves settle into a kind of discipline, moving the baby's body when it needs to be moved and—equally important—holding it still when stillness is required. Those responsive nerves adjust the baby's balance. They send hands gripping tightly if mother almost drops him, arms flailing outward if the infant feels off balance. The mother (or father or caretaker) may not think of herself as helping the baby while she busily hustles the child around on errands. But she is. Without such movements, the nerves aren't so pushed into making the needed connections. On multiple levels, it seems, the developing nervous system craves stimulation.

Left alone, the infant's body itself will compensate for stillness with a kind of self-stimulation. Thus the baby monkey, alone with his statue mother, has only himself to provide the necessary activity. Perhaps then, so the theory goes, it's a lack of motion that induces the rocking and the odd flapping of hands seen in the little monkeys raised with a cloth mom. The same self-directed behaviors are called "stereotypies" in children suffering from disorders such as autism. Those behaviors, too, seem to be linked to some stumble or unmet need in the developing nervous system. There are clearly other possibilities, obvious genetic ones, to induce such a stumble. Still, early research—such as Mason and Berkson's—and later research both suggest that even a little baby-rocking can be a very good thing and that the old rocking chair may be something that a doctor should prescribe.

At least, that was what Mason and Berkson started to consider as they watched their little monkeys respond happily to their restless surrogate mother. They also considered other, equally compelling possibilities. "It was a beautiful study but it was confounded by one thing," Berkson says now. "As this surrogate mother moved around the cage, the baby would swing and jump and do all sorts of activities that the others didn't do." A baby who had a statue-like mother could sometimes resemble a statue himself, but that didn't happen if the surrogates provided some action. Instead of just clinging and holding still, the babies seemed to respond, almost joyfully, by becoming active themselves.

The study thus leads to a circular kind of thought process. If motion is needed early in life to wire the nervous system correctly, who is the critical actor? Is it the mother, rocking the child? Or is it the child, moving in response to the mother's pacing and swinging? Or does it matter which? Perhaps, once again, this is all about interaction—motion by the mother induces responsive motion in the child. Bill Mason, now at UC-Davis, thinks it was the unexpected back-and-forth relationship that encouraged more healthy development in the little monkeys: "I've come to realize that the mobile surrogate was more like a real monkey than we had expected." Mason suspects that just the extra wiggle "provided a limited simulation of social interaction, which the stationary surrogate did not."

It wasn't just that the mobile surrogate wobbled here and there around the cage; it was that the baby never knew *how* it was going to wobble. The mother might swing left and right, back and forth. That made the surrogate an unexpected kind of companion after all, and gave the element of surprise in mothering that Mason thought mattered. The swinging surrogate could withdraw without warning; it could swing round without warning and gently bop the infant on the head; its comings and goings required attention and adjustment. The monkeys could space out with cloth mom; she wasn't going anywhere at any time. Cloth mom was completely predictable. You didn't have to think about what she might do or how best to approach her. More or less, you could climb her the way you climbed a tree. But the mobile surrogate required strategy from the baby monkeys. They had to pay attention to jump on and receive a little contact comfort. They chased, pounced, wrestled. She encouraged their interest in having fun by being fun. Rough-and-tumble play was about three times more frequent in monkeys whose mother was mobile than in those living with the cloth mom. Mason and Berkson hadn't expected their traveling parent to be quite so interesting, but, as Mason said, "We had unwittingly created a social substitute."

Of course, we all hope for more in childhood than the mechanics of good mothering. A warm wobbly stuffed dummy is hardly anyone's ideal of a mother. But what the mobile surrogate told Mason, at least, was that even a whisper of social interaction makes a difference when you happen to belong, as rhesus macaques do, and as we do, to a very social species. Evolution has not left us hopelessly vulnerable to the indifferent parent, the minimal mechanical mother. That a baby monkey can adapt relatively well to a swinging stuffed pillow of a mother is a reminder that we are designed for survival. We can, if need be, get by with remarkably little from the parent we have. Of course, we may only just get by.

Later, Mason would look further at the ways that a baby monkey can spin a social support network out of fragile threads. He came up with a wildly innovative, some also say peculiar, experiment.

It involved baby monkeys, hobbyhorses on wheels, and dogs. After all, dogs are also a social species and they and baby monkeys get along just fine in a buddy kind of way. They'll play together, sleep together, groom each other. But a female dog, caged with a small monkey, does not act in an especially protective way. The dog doesn't come running to a wail of monkey distress or break out in primate maternal behaviors. Mason calls dogs and monkeys "generalized" companions. His term is basically a scientific way of saying that dogs and monkeys can be friends. Still, a friend, even a shaggy member of another species, must be a whole lot better as a cagemate than a roll of terry cloth with a croquet-ball head. Friendship, by definition, is a give-and-take relationship. Mason decided to compare dog-raised monkeys with monkeys raised by an inanimate surrogate. But he still wanted something better than a plain old stationary cloth mom, and this led to the plastic hobbyhorses. They moved, or at least they rolled; and they were available for clinging. On each horse Mason carefully placed a softly padded saddle.

Six baby monkeys were placed with dogs; six with hobbyhorses. All the monkeys showed visible affection for their companions. They all grew up into competent animals. But they were very different in some very important ways. Mason watched these monkeys for four years, even after they had been moved into a bigger colony. Over and over, he found that the dog-raised monkeys were more engaged with other animals and more interested in the rest of the world.

In one test, he put the monkeys into a box with peepholes in the side. It was not like a Butler box. The monkeys didn't have to do anything to see out. There were peepholes, open and available for the looking. The animals just had to be interested in the possibilities outside. If they did put their eyes up to the peepholes, they'd see a picture. The dog-raised monkeys were simply fascinated by the chance to see something new. If Mason changed the picture, they would crowd the peephole and curiously study the differences. The other monkeys tended to hang back. They were a little nervous, a little uncertain about that outer world. If Mason gave the animals a puzzle to solve, the same pattern developed. It was the dog-befriended animals that exuberantly tackled the problem. The monkeys raised with the hobbyhorses tended to quit if the puzzle was too difficult. It worried them. They'd go back and comfort themselves by holding onto their hobbyhorse again.

The difference, Mason argues, is that wonderful underrated opportunity to interact heartily with your companion: fight, play, share food, hog the bed space—even the shoving match matters, the constant back and forth as one player influences the other. "Stationary surrogates and hobbyhorses surely provide few opportunities for the developing individual to experience the fact that his behavior has effects on the environment," Mason says. Maybe more important, if you successfully snag the last piece of cake, if your bedmate gives you that extra inch or so of space, you learn that you can, sometimes, exert control over your environment. "Inert mother substitutes make no demands, occasion no surprises," as Mason says, and thus teach us nothing about managing our surroundings and, occasionally, ourselves. You need not pay attention to them, but it is in paying attention to others that we acquire social skills, learn the "fabric of social interaction."

Harry also came to realize that a cloth mother's impenetrably passive nature made her a hopeless parent. She might be as cuddly as a fleece, but fluffy availability was never going to be enough. And it wasn't just that cloth mom didn't hug back. It was all the other things she didn't do: She didn't teach, direct, or steer the baby toward others. From cloth mom, the baby really learned more isolation, separation from others. "Growing up to be a monkey is an intricate process involving both ties of feeling toward other monkeys and the learning of monkey behavior patterns," Harry said in a newspaper interview. Real living breathing mothers, he added, were not yet dispensable. Not for monkeys—or for their human cousins.

Because we are social animals, it seems that one companion serves to connect us with another. And it's worth remembering that rhesus macaques are definitively social animals. In the wild, or in cages that are large enough, they spend their lives in big, tumbling, interactive troops. They play games, they schmooze each other, they groom each other. The females help care for each other's children; the males plot and form alliances, triumph together and sulk together, according to the results. A troop is strictly structured, organized by hierarchy, by social awareness, by street smarts about who's a friend and whose back really, really needs to be scratched.

By comparison, surrogate-reared animals are like alien monkeys from the planet nowhere. The cloth-mother-raised babies didn't engage in any of that all-important schmoozing. They didn't play with other monkeys; they didn't swing into the usual spring mating season. They had no idea what to do with other monkeys—as friends, as enemies, as potential mates, as casual companions. No one had showed them the social ropes and they simply couldn't find them without help. "The surrogate mother can meet the infant monkey's need for an object of affection," Harry said. "But it cannot teach the infant to groom itself or others, as the real mother does. Nor can it replace the mother and the other members of the monkey group, young and old, in providing the variety of social rules that the young monkey needs to make its way in the monkey world."

The original concept of his surrogate mother was still grounded in that sterilized notion of a healthy baby—clean, fed, warm, disease free, isolated from harm. Once again that had been proved incomplete at best and destructive at worst. "Harry originally thought he could be a better mother than the monkeys were," Levine says. "And he was wrong." Harry hadn't stepped all the way back to the complete Watsonian model of maternal indifference. He knew that touch and affection mattered. What he hadn't appreciated was that this matter of mothering was so complicated. But Harry was learning fast. He and his students continued in their attempts to dissect motherhood, not just what made a good parent, but what elements made the cloth mother such a bad one. What exactly were they seeing in this collapse of the surrogate-raised monkeys? Was it an insecure attachment? A failure of maternal responsiveness? A lack of social education?

The Wisconsin researchers would reach an all-of-the-above kind of answer. Love, Harry would eventually argue, was not built of one relationship but many. Our love lives, all of them, forge links in a healthy chain of normal development: maternal love, infant love, paternal love, friendship, partnership—one connecting to the next and then the next. The early attachment is the first link of that chain, the start of our ability to connect with others.

Now cloth mom wasn't all awful. She was always there and she was never rejecting, much as any mother of a very small child needs to be. Becoming a parent means that patience becomes an ever-elastic attribute, stretching farther and farther as needed—and that will be much farther than a novice parent first appreciates. No matter how much a baby wakes up in the night or throws up all over her mother or screams in her father's face, most of us know that our job is to answer the cry, clean up the mess, comfort away the scream without anger. We learn to walk away, to take our exhausted frustration out on, preferably, the nearest inanimate object. And babies all need that rock-solid acceptance as well, Harry insisted. Even monkeys know this. Rhesus mothers almost never punish a baby monkey in the first three months of his life, no matter how he tugs or pulls or makes his mother's life uncomfortable. Cloth mother was perfect in meeting this particular challenge. She never slapped, never rejected.

And yet, Harry also came to believe that one of the most important things that the mother must do, not at first, but soon enough, is to nudge the child away. There were two problems, as he saw it, with good old cloth mom. One was that she didn't groom, or talk, or make faces, or directly and indirectly cue the baby for relationships with others. And the other was that she didn't cut him loose to engage in those relationships. The mother needs, absolutely, to be there for baby, but she needs to show him how to be there for others. In a social species, Harry said, one relationship is never enough. We build a world of connections. We weave them—contacts and friendships and family and loves—into something that we lightly call "a support network," and which is really the safety net that catches us as we balance our way along the high wire act of every day life and from which all of us occasionally fall.

The one thing that made cloth mom sound so appealing at first—her never pushing baby away—turned out to be one of her eventual liabilities. If you turn that passive acceptance around, it meant she never encouraged her child to let go, never gave him—you may see this coming—that gentle push out into the rest of the world. In this light, Harry reconsidered Konrad Lorenz and his adoring flock of goslings. Lorenz's famous imprinting work with the greylag geese had shown that the youngsters were dedicated to the mother who first hovered over them. If the goose wasn't there (having been removed by the scientists) and Lorenz was, well then, the goslings followed Lorenz with compulsive dedication. There are still wonderful photos of Lorenz, upright and gray-bearded, marching through a meadow, trailing goslings behind him like beads on a string. Lorenz called this dedicated behavior "imprinting," suggesting that the mother is imprinted, like words set into stone, into the baby's consciousness. John Bowlby was a friend of Lorenz's and an admirer of his work; that hard-wired connection between mother and child was a touchstone piece of evidence in the building of attachment theory.

And yet, at some point, a gosling must be able to stretch that connection. Eventually, waddling, flying, climbing, and walking away from your mother is also a survival instinct. Even a bird needs to grow up, find its own mate, build a nest, and raise its own family. Or, in Harry's still poetic view of it:

How does an infant break away
To be a goose himself some day?

Psychologists had long considered that all of us—goose or child—require some independence. Babies need total acceptance, but as they grow, too close and cuddly a nest is not necessarily a good thing. New York psychiatrist David Levy, whose work would eventually help support attachment theory, also spent time trying to determine when a child should stand on his own. His 1943 book, *Maternal Overprotection,* sometimes reads like an ominous Brothers Grimm tale of mothers who hedge their children in and surround them, like that wall of thorns around Sleeping Beauty's mythical castle. Levy's book contains twenty case studies, every one a lesson in the destructive effects of denying your child room to grow. Levy tracked one gentle sixteen-year-old boy whose mother always went to the movies with him and explained all the action so that his mind wouldn't be "poisoned" by the wrong ideas. Another mother told Levy candidly that she hoped to keep her son "her baby" until he was at least thirty-five.

Some of those homebound children simply obeyed maternal orders, growing ever more withdrawn and deferential. Other children beat against the walls of their cells. One fourteen-year-old boy, whose mother wouldn't let him play with other children, or play chess, or even read detective novels, took to deliberately tracking mud through the house, cutting holes in his clothes, and screaming at his parents and siblings. All twenty of the children in his study, Levy noted, had extraordinary difficulty in making friends. They were loners, and Levy believed their mothers had turned them into misfits.

The question for Harry Harlow was whether the ever-available cloth mom exerted a similar warping influence. She wasn't a perfect parallel to Levy's suffocating parents. She never held the children back, trapped them with her. They could leave whenever they wanted. She just never encouraged them to take those first steps away.

Harry did eventually see some parallels with Levy's overprotecting mothers. Cloth mom, he said, by being so passively acceptant never nudges her charges toward other relationships and "thereby never encourages independence in her infant and [affectionate] relationships with other infants and children. The normal state of complete dependency is prolonged until it hinders normal development." The question still remained, of course, of how exactly to promote healthy independence. Should mother or child let go first? Should the mother softly hint the child toward others? Or does the child need to push for herself, open those still-fuzzy gosling wings, tread air for a while? And when? Is there a right time to fly, an equally right time to be gathered back into the nest?

Everything out of Harry's lab, and elsewhere in primate research, says this is a delicate moment of negotiation, all timing and small steps. Robert Hinde, Bowlby's good colleague at Cambridge, saw it almost as a dance: When a child is very young, the mother works hard to pull the child close for her own safety. But as the child grows, is less vulnerable, the mother becomes a little less protective. She takes a step or so back. Then the baby's first response is to move toward mother. Hinde reported that small monkeys would begin calling more, nestling tighter. If the mother kept pulling back—mother monkeys would now cuff an older child who pulled too hard on her fur—the youngster would learn to seek comfort elsewhere. She might scoot over to other young animals, play longer away from home. Eventually, she would have strengthened those friendly relationships as well, broadening her base of support.

The bond between the specific mother and baby, though, provided the background music to this dance. If the mother was never very supportive, what you might call a John B. Watson–approved monkey who pushed the baby away from the beginning—then the little infant didn't want to go elsewhere. The young monkey didn't feel secure enough to risk the next relationship. Harry began to see a sequence in this, a surprisingly strict order. First, the baby needed to believe solidly in that relationship with his mother, to be securely attached. If he didn't feel that kind of solid support, it was much harder for him to turn easily to other animals. There was an obvious common sense conclusion, Harry thought, to the construction of this chain of relationships. If the first one failed you, it was much harder to forge the next. So the little monkeys stayed close to home, trying to mend their link to mother, failing to build the links to others.

Leonard Rosenblum, after he had graduated from Wisconsin, did some studies that beautifully illuminated this idea. He used another species of monkey—rhesus are not the only macaque species, after all, and cousins of all shapes surround them: crab-eating macaques, bonnet macaques, pigtail macaques, lion-tailed macaques, Barbary macaques, Japanese macaques. Macaques in shades of gray and brown and silver-gray and flaming gold, tree-climbers, water lovers, foragers, homebodies. Sweet-tempered macaques and evil-natured ones, good mothers and indifferent. Pigtail macaques tend to be exceptionally affectionate mothers. Bonnets are far less focused in their offspring. And what Rosenblum found was that pigtail macaque babies, cuddled and fussed over and protected, found it easier to move on to other relationships. But the bonnet babies were continually trying to repair to the home front. They were clingier. Instead of the independence you might have once expected from being pushed outward, they hovered near home longer. So here you had an apparent paradox: To create an independent child, you needed to allow the baby to be dependent.

And then you had to know when to let that baby go. At one point, recognizing cloth mom's failures, Harry wondered whether pairing baby monkeys together would answer their need for a real relationship. But babies, remember, are made to cling and hold on and gain comfort from a security figure. So what the scientists ended up with was not two secure monkeys but two monkeys who wouldn't let go of each other. Harry explained it like this: "The clinging together for contact comfort overwhelms the babies. They don't know when to stop . . . they haven't even enough sense without mother to know when to start playing."

The infants wouldn't reject each other at any stage of development. They were almost worse than cloth mom. At least she didn't keep a stranglehold on her baby. "Harry discovered that if you rear two infants together, it's almost as bad as total isolation," says Jim Sackett, now at the University of Washington in Seattle. "They develop a tight clinging behavior and it looks cute, but if you separate them they go to pieces. It's deadly for later socialization. Nobody in their right mind, who knew Harry's work, would raise rhesus babies in pairs. Adults in pairs, yes. Infants, no."

So the list of mother mechanics gets longer here: warmth, motion, affection, and now enough sense to know when to hold tight and when to nudge a child away. Psychologist Irwin Bernstein, at the University of Georgia, calls such behaviors—a sense of timing in relationships, an awareness of when to hold tight and when to let go—"social intelligence"; he notes that back in the 1960s, when Harry first started talking about this, it was not a topic on psychology's radar screen. "It was Harry's genius to recognize that the baby monkeys were abnormal emotionally and that it was 'social intelligence' that they lacked," Bernstein says. "This was not an area much investigated in the middle of the twentieth century."

The mother-child bond, on the other hand, now had everyone's attention. Psychologists were riveted by the notion of mother love, contact comfort, and attachment theory. Now, as Steve Suomi dryly notes, there was a "relative preoccupation with mother-infant relationships by those in the mainstream of child psychology and psychiatry." Thus the explosion of work that so inundated the editors of *The Competent Infant.* It was going to take a little time for psychologists, in general, to see the mother-child relationship in the wide-field context of "social intelligence."

But, as Suomi points out, a monkey colony almost forces the broader panorama of relationships upon you. The whole community is there, in effect, and if you put mothers and infants, juveniles and other adults together you see a society a tumble with small monkeys knocking each other over, chasing, exploring, arguing, zipping back and forth from friends to family to friends. There was no way to watch monkeys and believe that one relationship alone was enough.

You needed social skills on a far bigger scale to survive. For one thing, Harry said, "Monkey groups can spot a stranger a mile away, and if the stranger does not recognize its predicament and display the appropriate submissive behaviors, it is almost certain to be threatened, repeatedly attacked, repelled and perhaps even killed. Knowing one's friends has enormous adaptive import, even among nonhuman primates."

In rhesus society, with its rigid top-to-bottom hierarchy, knowing your friends, their place, your place, adds up to a basic formula for survival. No one is born with that knowledge, and yet, from very early on, a child needs to know where he fits. If social intelligence has to be taught, the suggestion is that every child—human or monkey—requires a dedicated teacher. "Monkeys are not honeybees," said Harry, and preprogrammed responses are not going to get them through life. "The best rhesus monkey genes in the world do not guarantee that the individual possessing them will be socially competent," as Suomi puts it. It's that absolute requirement for social intelligence that also helps explain why cloth mother—who knew nothing and could teach nothing—in the end turned out to be such a bad surrogate.

When mothers delicately shift their children into other relationships, they also shift them into new levels of social learning. As Harry pointed out, there's a simple name for the next phase in building relationships. It's called play and it's one reason why it is so important that parents encourage their children to form friendships with peers. When does play with peers become a major part of social life? In monkeys, it begins at about three or four months—comparable to a human toddler of two years. Watching the lively macaques, it was clear to Harry that to play well, a whole new set of social skills was required. Game playing between peers, obviously, doesn't look much like a mother-child relationship. But in the Wisconsin lab then—and in experiments that continue today—observers were often startled by how closely childhood play could resemble adult interactions. The parallel was strong enough, as

Harry speculated, to suggest that play is a kind of "prototype" for adult interactions, a test run for the future.

Monkey play at his lab tended to go in two primary directions. One was "rough-and-tumble play"—what Harry described as a monkey wrestling match—with lots of rolling and scraping and almost no injury. This is between friends, after all, so no one pounds all that hard. The alternate favorite game was something similar to what we call "tag" and scientists call "approach-avoidance play"—chasing, running away, very little actual physical contact. At one level of observation, you could watch monkeys do this and see a wonderfully rowdy time. At another level, you could see a really terrific way to pick up a few fundamental life skills.

In rough-and-tumble play, a child learns judgment—how far to push without getting hurt, who is going to take the game too seriously, when to back off, when to push forward. Tag, too, lets you judge speed, interest, who's going to run, who might decide to stand back, who's a good sport or a bad loser. And both games teach you another equally important life lesson: pleasure in the company of others.

As the monkeys grow older and play harder, they get better at sending and at reading the kind of messages that we call nonverbal communication. Peers tend to reinforce behaviors—reciprocating when they like an activity, ignoring or turning away when they don't. So during play, you can also learn what makes your friends leave and how to coax them back. Most of us want them back. Rhesus macaques and their human cousins aren't built to be loners. Reconciliation, among other skills, matters. We do best, live longest and happiest, when the social net stretches firmly beneath us and we, in turn, serve as strands in the adjoining nets that protect our friends and family.

The Harlow lab put the idea of a complex social network to the test in a study devised by Peggy Harlow. In the midst of the mother love studies, Peggy had been thinking about family itself, the basic support system of home—mother, father, and children, all together. To do that with the rhesus—not a monogamous species in any way—

she had to find a way to keep the father at home. Her "nuclear family apparatus" had nothing to do with natural rhesus society, in which females care for the young and fathers move on to other, less needy company. Peggy's apparatus was designed to ask something of rhesus macaques that they never gave—permanence, togetherness in parenting, a stable home, and what Harry jokingly called "blissful monogamy."

The design was for a neighborhood of four macaque families. The Wisconsin lab workers converted an attic of the primate lab into this community. Each cage-house sheltered a mother, a father, and their children. Each house opened onto a central playground equipped with climbing ladders, swings, and toys. From the playground, then, you could visit any of the neighbors—if you were a child. All the doors were the right size for small monkeys only. Mom and Dad couldn't get through. The young monkeys had the run of the place. They lived in a child's neighborhood. The little animals could play together, they could hang out at a friend's house, they could scamper for home if alarmed or tired. Their parents would always be home.

The Wisconsin psychologists worried that the big males, trapped with their children and only one mate, would turn mean and abusive. But the monkeys surprised everyone. It turned out that even an arrogant alpha male macaque could find untapped potential in the right circumstances. "What was really astonishing was that the males then took part in protection and rearing," says Gerry Ruppenthal, now at the University of Pittsburgh, who worked with Peggy on the nuclear family project. "At that point, the thought was that a male rhesus didn't interact kindly with anyone. And she proved them wrong."

The big males anxiously guarded the smaller animals. At home, they played with them with surprising cheerfulness. They tolerated those annoying kid behaviors—pinching, biting, and tail and ear pulling—that might have angered them in the more free-ranging colonies. In another life, the burly male monkeys might have knocked the baby monkeys far, far away. But in this life, they often settled for a shrug, the monkey equivalent of rolling their eyes.

The young monkeys thrived in this lively community of family and friend. Harry would remark that of all the animals in his laboratory, these were the most confident, the most socially adept, the most outgoing—and, surprisingly, the smartest. Monkeys raised in the nuclear family community were faster and more accurate on the most difficult tests of the WGTA. Their minds seemed sharper and more flexible, as if learning to handle a multitude of social relationships had built their brains to handle other challenges well, too.

Can one strong relationship ever substitute for such social complexity?

In one later test of this, Harry and his students decided to push the limits of a single-relationship life. They didn't even consider cloth mom for this study. They kept the baby monkeys with their living, breathing, interactive biological mothers. The catch was this: Mother and baby were separated from the rest of the colony. They had no one else for six months; there was no chance for that lively interaction with playmates. The offspring grew up obsessively and abnormally shy. Even when surrounded by other young monkeys later, the single-relationship monkeys avoided them. They didn't want to play. They rarely groomed and befriended their new companions. When approached, they tended to be hysterically defensive; in the terminology of the lab, hyperaggressive. "In short," Harry reported dryly, "they do not make good playmates." And the longer the baby was isolated with his mother, the more inept he became at making friends—not unlike David Levy's overprotected children back in the 1930s.

Harry's lab would prove the paradoxical nature of love. The one, the only relationship, isn't enough. And yet, the one, the first relationship can be everything, swamp everything. That the first attachment, interactive relationship, social connection, call it what you will, is so potent has less to do with mother than with child. Joseph Stone's description of a baby's intense attraction to a parent was addiction—and that may well be exactly the word, although no scientific term really does justice to the blinding, white-light intensity of commitment that a young child will give to a parent.

In their exploration of love, in all its shapes, Harry and his students would illustrate this precisely—and painfully. Joseph Stone also would include these latter findings in *The Competent Infant* as an essential part of the baby story. These studies, as well as the gentler ones, helped transform the landscape of child development research, and they remain essential. Still, it is genuinely hard, even now, to read these particular testaments to baby love. Even in small black print on the yellowing pages of fading scientific reprints, they make all too real the wholehearted nature of what a child gives a parent. In their give-everything vulnerability, they are yet word-by-word painful to read.

The work in question didn't begin as a test of commitment but as an experiment to investigate the effect of a pathological mother on her child. To do the study, the lab team built what Harry called evil or "monster" mothers. There were four of them and they were cloth moms gone crazy. All of them had a soft-centered body for cuddling. But they were, all of them, booby traps. One was a "shaking" mother who rocked so violently that, Harry said, the teeth and bones of the infant chattered in unison. The second was an air-blast mother. She blew compressed air against the infant with such force that the baby looked, Harry said, as if it would be denuded. The third had an embedded steel frame that, on schedule or demand, would fling forward and hurl the infant monkey off the mother's body. The fourth monster mother had brass spikes (blunt-tipped) tucked into her chest; these would suddenly, unexpectedly push against the clinging child.

And what did the babies do when all this happened? If possible, they clung tighter. At least, that was the response of the little monkeys with the airblast and the shaking mothers. The other monster mothers could successfully remove the infants by the force of the spikes or by literally throwing them off. But those baby animals, as soon as they safely could, returned and hugged their mothers again. Time after time, the babies came back, Harry wrote, "expressing faith and love as if all were forgiven." The experiment did not, in fact, create psychopathic monkeys. It created neurotic ones, yes, but not crazy ones. Its primary finding was entirely different from the expected result: "No experiment could have better demonstrated the power of any contact-comfort-giving mother to provide solace and security to her infant."

Or, to turn it around, no experiment could have better demonstrated the depth and strength of a baby's addiction for her parent. Or how terrifyingly vulnerable that addiction makes a child. These little monkeys would be frightened away by brass spike mom—and yet it was she they turned to for comfort. They had to; she was what they had. Here indeed was further evidence of that haven-of-security effect, for better and for worse. It doesn't always keep you safe. If your mother is your only source of comfort and your mother is evil, what choices are left you in seeking safe harbor? No choice except to keep trying to cast anchor in the only harbor available.

Harry and his team would find the same pattern when real mother monkeys were rejecting or abusive. The scientists marveled at "the desperate efforts the babies made to contact their mothers. No matter how abusive the mothers were, the babies persisted in returning." They returned more often, they reached and clung and coaxed far more frequently than the children of normal mothers. The infants were so preoccupied with engaging their mothers that they had little energy for friends. The clinging babies' energy was directed into their attempts to coax a little affection out at home. Sometimes the real monkey mothers did respond, gradually, more kindly. But while trying to reach mother, the little monkeys never had time to reach anyone else. "Like most human children, young monkeys play poorly when they are frightened, unhappy, or preoccupied with their mothers' activities," says Suomi. And playing poorly meant that the other monkeys didn't always want to play with them. It was a typical childhood reaction: You aren't any fun; you might as well go home and be with that mother of yours.

In one of the Harlow essays, included in *The Competent Infant,* Harry addresses this issue in the most scientific way: "All infants show the filial affectional system, and they all show it in the same way. On the other hand, by any measure of maternal behavior, the maternal affectional system shows high variation." Translated to everyday language, the first part of that statement means that almost all parents are guaranteed that their babies will love them. The second part is a reminder that the baby has no guarantee at all of being loved in return. Love, as science so directly reminds us, as Harry and his studies illustrated with such knife-edged precision, can never be taken for granted—not on the first day we draw breath, not ever.

EIGHT

The Baby in the Box

When one accepts propinquity / instead of chilling dignity / a life becomes depression free / as every life should always be.

Harry F. Harlow, undated

When Harry first began exploring love, there was a leaping sense of joyfulness to his discoveries. This was wondrous, amazing science, and even the lead researcher himself wasn't immune to its dazzle. When the barely submerged poet in him broke into verse, he composed light-hearted, even goofy odes to mother-child love across the animal kingdom.

Like this one:

This is the skin some babies feel
Replete with hippo love appeal
Each contact, cuddle, push and shove
Elicits tons of baby love.

And "The Elephant":

Though mother may be short on arms
Her skin is full of warmth and charms

And mother's touch on baby's skin
Endears the heart that beats within.

But now, after watching and tweaking relationships, he was thinking less about endearing love and more about love that must be endured. What is the other side of glowing affection, the opposite of the tender touch? There was no inclination to pen doggerel this time. The questions were darker and more dangerous, and, for a scientist, more risky. It's one thing to study the necessity of love and touch and how to support a child. It's another to consider what might happen if you destroy that support system. And yet, can you understand love without understanding hurt? Harry was near enough to that question, close enough to see promise and the peril. Close enough to know that another question was implicit—was he willing to risk his shiny reputation to go there, into the troubling country of love and pain?

And yet he thought that journey mattered, maybe a lot. There were these tantalizing suggestions that even the "best families," even the tight-knit community of the nuclear family monkeys, could not entirely depend on the kindnesses of love. Lorna Smith, Harry's graduate student and a participant in some of the early mother love studies, was drawn into the fascination of families—how they worked, how they didn't. Now Lorna Smith Benjamin, she holds a professorship at the University of Utah and specializes in working with dysfunctional families and helping repair the effects of traumatic childhoods.

She remembers, as a student, just watching the rhesus families at work and play. And she also remembers an evening with Peggy Harlow and the nuclear family study: "I'll never forget. Peggy was making a narrative of what she was looking at and what she was thinking. It was like a little suburban neighborhood with the kids out playing and the parents at home. They had lights set up that would come on in the morning and go out at night, for the day-night cycle. When the big lights went out, the room had dim backlights so that the researchers could still watch."

While the two women were watching, night fell over the little neighborhood. "She told me to watch one particular baby. When lab lights went off, everyone went home from the playground to snuggle up with mom and dad and go to sleep. This one little character was still out in the playground. She said, 'He'll go home, but not until his parents are asleep.' After twenty minutes or so, the little thing crawled into his cage, found a patch of fur to cuddle against and went to sleep. And she said, 'He has abusive parents.'" The little monkey, Peggy said, coped with his parents by staying away from them as much as possible. And in the dark, when his mother and father were harmless, he came home.

The nuclear family studies were a reminder that even in the nicest neighborhoods, families don't always work. To be fair, the studies mostly illuminated the ways that healthy families balance the tensions of multiple relationships and demands. Harry had expected that with each new baby, the mother would become dismissive of the older siblings, too busy with the little one to bother with the others. And, in fact, directly after the birth, monkey mothers did tend to push their older children away. The mother would obsess over the baby, hold it tight against her.

At first, she would turn away from the older children. They couldn't accept it. Wasn't she also their mother? They would sneak up to mom and cuddle at night while she slept. They would lean against her, cooing and appealing, during the daytime for as long as she allowed it. They would woo her back. It was usually no more than a night or two before big brother and big sister were safe in the family huddle. The researchers had wondered whether the older siblings would slap the babies around, take out some of their frustrations. "Much to our surprise, the displaced infants did not overtly exhibit punitive signs of jealousy toward the newcomers," Harry wrote. "Although one male juvenile did engage in teasing his little sister at every opportunity when mother was not looking."

But even watching the most loving mother surrounded by her offspring, you couldn't miss how needy and demanding and vulnerable the smallest monkeys were. They were as unnervingly fragile as their human counterparts. Peggy watched them and she worried over them. She anguished over the monkeys when they became sick, worried about the fate of one whose mother had fallen ill and died. Shortly after losing her mother, the daughter became ill, suffering from a painful intestinal bloating. "Most people think that's due to stress," Ruppenthal says. "Dr. Harlow came in and she kept watching over that little sick monkey and she would bring it special treats. But it died anyway. She came in the next morning and she had tears in her eyes. She said, 'Gerry, we've killed her with kindness.' They were like little kids to her. She loved them."

The nuclear family studies were—deservedly—gaining Peggy real respect; she was beginning to regain the professional momentum she had lost when the University of Wisconsin took her first job away. She was now conducting research in the primate lab. She had a position as lecturer in the neighboring department of educational psychology. And she was beginning to prove, scientifically, something she believed in wholeheartedly. Her experiments were directed toward the idea that the whole family matters, that mother love works best in a communal sense, that it requires the help of father and friends and even the neighborhood. "When my wife became pregnant, [Peggy] talked to me about having kids, and gave me a lot of good advice about child rearing, the importance of interacting, parenting, a stable living environment," Ruppenthal says. "She was looking at that in the lab, sure. But her real perspective was the human perspective."

Peggy was still reserved enough, brisk and cool enough, that many of Harry's graduate students didn't see that human aspect. She was shy enough that she rarely shared such motherly advice. Ruppenthal knows that she was more widely disliked than not, that most students thought of her as cold to the heart. Ruppenthal doesn't care: "She was, absolutely, a wonderful person." And he still admires her research. "I think her hopes were that creating a stable compound environment would bring out the best in the animals. It showed that you can become a very sophisticated animal in a warm environment," he says. "It was far greater than she expected; it blew her away and it blew me away."

One can wonder where her fascination with family would have taken Peggy Harlow, whether she might, herself, have become a psychology star at the University of Wisconsin if she had had a little more time. But she didn't have the time, not enough of it, anyway. In 1967, she was diagnosed with breast cancer, already spread just beyond the bounds of control. Her illness would slowly, but relentlessly, help push Harry closer to his questions about the risky side of love. It would make real all those troubling questions of fear and loss and vulnerability that hovered around edges of relationship research. Harry would consider, once again, the dark places that love can lead you.

Peggy was still working like a foot soldier; if she was well enough to stand, she was at the lab, fussing over her monkeys, taking conscientious notes. Harry was still soldiering on, too, but he was desperately worried and was stretching to snap point. He was so exhausted and distracted that he could sometimes hardly remember why he was there, much less anyone else. "I saw him in the lab one Saturday morning," Ruppenthal says, "and he said, 'Who the hell do you work for?'"

Harry was traveling constantly, having confounded another of Terman's expectations and become a nationally sought-after speaker. In the main psychology department, his reputation as the least visible member of the Goon Park community continued. "The comment was, there's the East Coast Harlow, who lives at Kennedy Airport, and the Washington Harlow, who lives at NIH [National Institutes of Health] getting money, and there's the Wisconsin Harlow, who's never there," recalls Gerald Wasserman, a UW psychologist at the time.

Harry was no longer responsible only for the primate lab. The NIH had decided that since primate research was so promising, it would create a series of centers, spread across the country, to explore the scientific possibilities raised by our primate cousins. Harry Harlow was among the psychologists and physicians and primatologists who helped persuade NIH to invest in primate research. He also convinced the agency to name Wisconsin one of its seven regional primate research centers. Madison thus became the home of the only NIH primate center in the Midwest. It was an enormous honor—and an enormous added responsibility. Harry was directing the center, running his own lab, attending assorted committee meetings, and, when he could, pursuing research. Wasserman, now a behavioral science researcher at Purdue University, still remembers the passing blur of Harry Harlow in action: "He'd be at his desk and it would be piled high and he'd be carrying on a coherent conversation while he was opening envelopes and reading things and grabbing things. It was as if he could operate with two different minds at the same time."

Of course, no one maintains the two-brain illusion forever. Harry was stretching thinner, the proverbial rubber band, pulled by guilt and worry in one direction and by his need to prove himself, always, in the other.

What snapped him was another success. He'd always struggled with achievements, always worried that they signaled the last peak moment in his career. It used to confound his friends how much an honor would trigger Harry's insecurities. Gig Levine was among those who joined Harry in celebrating after the famed "Nature of Love" talk. Levine's strongest memory of that party, though, is not a jubilant one. It's of Harry Harlow huddled in the corner of a bar, sliding down into bourbon and self-doubt. The evening had begun joyously enough with celebratory drinks, and then more drinks, and then more. "And it wasn't just a drunk but a really black drunk," Levine says. "His mood just got blacker and blacker and he said to me, 'What am I going to do next?'"

In 1967, Harry Harlow became the first (and only) primatologist to win the National Medal of Science. He was called to the White House for a ceremony presided over by President Lyndon Johnson. It seemed to give him no pleasure at all; he told friends only about Johnson's impatience with the whole affair. Harry came home convinced that this time indeed there was no next, that with the medal he had topped out professionally. And in his personal life, although he wrote to friends about the promise of chemotherapy and about doing the best they could, Harry was preparing to lose his wife.

Steve Suomi, who now heads NIH's Laboratory of Comparative Ethology, remembers his arrival at Wisconsin for graduate school in the winter of 1968. Suomi showed up at the lab expecting to pursue primate research with the most famous psychologist in the department. Within two weeks, Harry had said to him, "I'm not doing well," and left for the Mayo Clinic, in Rochester, Minnesota, where he sought treatment of a paralyzing clinical depression. He would stay there for two months. The depression proved stubborn enough that doctors would move from drugs to electroshock therapy in trying to control the illness. The depression would moderate, but it would linger long after he came back.

Harry returned to Wisconsin a quiet man. He was silent about the time in Minnesota, withdrawn from the lab for one of the few times in his life, preoccupied just with making it through the days. He seemed to be a man worn out with research, a man who had finally given up on the next big project. "People would talk about what he used to be like," says one former graduate student. But, in reality, he was also considering a new research direction, a next challenge. And this one, finally, would move him over the line. He was thinking now about the darkest side of love, not what it gives but what it takes away. He wanted to create a monkey model of depression. He wanted to explore the biochemistry of this particular wasteland. If they could find a good way to study it, he was sure there were better ways of helping people lost in the Arctic zone of depression. He could serve as a personal witness to how much that work was needed.

Harry had a clear image of what that model of depression would be. It would be complete and utter aloneness; isolation taken to the icy extreme. Years later, he would look back on those experiments and group them together in a book chapter titled "The Hell of Loneliness." We all live, Harry wrote, with periods of social isolation: illness in the family, leaving familiar friends and family, business trips, going to college, divorce, the death of someone we love. "The extremes of human social isolation might be, at one extreme, a child's first day at school and, at the other, the solitary confinement of a criminal offender. The strangeness of a child's first day at nursery school, kindergarten or first grade, after mother leaves, will usually dissipate after the first few days among socially raised tots."

But if it doesn't fade away, if we don't connect, if we feel trapped in solitude, well, all of us know just how painful that really is. "The total social isolation of solitary incarceration is considered so drastic that Americans pride themselves on reserving it for the most pernicious prisoners," Harry wrote.

That knowledge had been simmering at the Wisconsin lab for years. Shortly after the first cloth mother studies, John Bowlby had come to meet with Harry and tour the facility. The rejecting surrogate studies were underway then, the monster mothers designed to push the babies away. The point of those experiments was to see if rejection induced psychopathic behavior. And it hadn't, the baby monkeys just kept coming back, trying to tighten the relationship, make it better.

Bowlby had been consoling about the apparent failure of the rejection study. No one has a winner every time, he reminded Harry. And then Bowlby went on his tour of the laboratory, where most of the animals were caged alone, according to the practice of the time. As Harry's students remember it, John Bowlby came back shaking his head. "Harry, I don't know what your problem is," he said. "I've seen more psychopathy in those single cages than I've seen anywhere else on the face of the earth." The monkeys were sucking themselves, rocking back and forth, cuddling their own bodies. "You've got some crazy animals," Bowlby said. In later years, Harry would laugh about Bowlby's ability to see what he himself had been blind to. "It takes a psychiatrist to have a psychosis," he said.

The first paper, focusing on the effects of isolation, was published out of Harry's lab in 1960. Bill Mason was the primary investigator on that study. In the same way that they had tried to take apart the mechanics of mothering, the Wisconsin researchers tried to explore exactly what made isolation so destructive. Was it the loss of physical contact only? What if the monkeys also couldn't hear any other animals? What if they couldn't see a single companion? They tried soundproofed cages, cages with solid walls that allowed no view of another animal. But it was difficult, maybe impossible, to filter out those separate effects. Because isolation just hammered the monkeys, flattened them out.

A rhesus macaque could make it with one relationship, even a swinging surrogate, a dog. But he could not make it alone. The effects of isolation—the despairing huddling—could look a lot like depression. Both Rene Spitz and John Bowlby had written about the way infants seemed to tumble down psychologically when they were separated from their mothers. Spitz called the numbed apathy that he observed an "anaclitic depression." He charted its progress like this: first, protest (symptoms: screaming, tantrums, weeping); and then depression (symptoms: withdrawal, slowness of movement, stupor).

The unanswered scientific question was whether this response to separation was true depression. Not every baby tumbled so simply into apathy. If the child had a restrictive mother—one who was continually confining her, like David Levy's overprotective moms—then, maybe not surprisingly, the child didn't seem to miss her mother quite so severely.

What Harry and his students worked out, then, might strike you as pure common sense. But, again, that was always Harry's measure of good science. The children who really suffered, the little monkeys who wholly grieved, were the ones who felt that they had genuinely lost something. "In other words," Harry wrote, "depression results from social separation when the subject loses something of significance, has nothing with which to replace that loss, and is incapable of altering this predicament by its own actions."

So what if you created just that scenario—total loss, total isolation, and total helplessness? If love is necessary to health and happiness, what happens if you strip a life completely bare of affection and connection? Wouldn't you then expect the kind of crippling despair that sends grown men off to clinics to be shocked and shaken back into a functional existence? What are the costs of belonging to a species that can never quite go it alone? How much can we actually bear? Everyone can take some loss and some loneliness, but there seems to be a point, different for each, when the burden becomes too much. One of the hallmarks of depression seems to be the crossing into that place where helplessness overwhelms almost every other sensation. If you want to accomplish despair in a laboratory, then, where do you begin to find that point of no return?

The first isolation experiments, of course, weren't looking for depression as an end point. They were pure explorations into the power of loneliness. The closed-off cage was an example. It was a blank space, equipped with a one-way mirror. The scientists could look in but the monkey inside could not see out. He had no company but himself. A baby monkey could be raised, almost from birth, without seeing anything except the experimenter's hands as they changed bedding or put in fresh food and water. The researchers placed a few infant monkeys into these boxes for thirty days. When the monkeys were moved, they were so "enormously disturbed" that two of them refused to eat and starved themselves to death. After that, the scientists at the Wisconsin lab force-fed monkeys coming out of isolation, to make sure the animals stayed alive.

The next experiments isolated baby monkeys for six months, and the next for an entire year. If the researchers kept a monkey in isolation for twelve months, they ended up with a rhesus macaque entirely new in the natural world, an animal who didn't explore, didn't play, barely moved, appeared alive only by the thud of its heart and the sigh of its lungs. Harry's students eventually had to re-isolate some of those animals. The monkeys were like born targets, so fearful, so helpless that they brought out the worst in their new companions. The other macaques would form a bullying ring; the isolates would cower within. "And as soon as the other animals would let up, these isolates would take off, which is a stimulus for more attack, and so you'd sit there and say to yourself, 'Please don't move. Please don't move,'" Bob Zimmermann remembers.

If the standard housing—one monkey to a cage—produced self-destructive behaviors, total isolation created far worse ones. Here, indeed, was psychopathology. These semiparalyzed monkeys, not surprisingly, were incapable of having normal sexual relations—of having any relations at all. When the lab crew had figured out a way to strap the dysfunctional females into a "receptive" position, they managed to induce a few pregnancies in already unstable monkeys. The result was an extreme reminder of just how dangerous an animal who has no "social intelligence" can be. "Not even in our most devious dreams could we have designed a surrogate as evil as these real monkey mothers," wrote Harry. "These monkey mothers that had never experienced love of any kind were devoid of love for infants, a lack of feeling unfortunately shared by all too many human counterparts." Most of the loveless mothers just ignored their infants. Unfortunately, not all did. One held her infant's face to the floor and chewed off his feet and fingers. Another took her baby's head in her mouth and crushed it. That was the end of the forced pregnancies.

So, Harry had evil mothers. He had crazy monkeys. He had unhappy and socially bizarre youngsters. But he still didn't see classic depression, that undeniable slump into misery. The researchers in his lab had created grief and loneliness and misery. But Harry was still looking for something more definitive, that paralyzing sense of life's being just too much, that state of being when air itself can feel as weighted as stone. "Depression in humans has been characterized as a state of helplessness and hopelessness, sunken in a well of despair," Harry explained. And that's what still seemed to elude him, that slide down into the bottom of the pit. Perhaps, he thought, they hadn't yet made their monkeys feel helpless enough.

This idea that depression springs partly from a sense of being trapped—a prisoner who has no escape—was just beginning to surface in psychology. While working with rats, scientists had found that if they exposed the rodents to inescapable electric shocks, so that no matter what they did the rats could not get away from that unpleasant jolt, the animals would visibly give up. The researchers could watch the rats collapse in what looked like a furry heap of despair. Later, in the 1970s, clinical psychologist Martin Seligman would begin developing such reactions into a theory of "learned helplessness" and the way that being stripped of power—or seeing yourself as so powerless—infiltrates every response. Seligman would come to believe that learned helplessness can drive not only depression but also the angry, lost behaviors often associated with it. He would also develop this understanding into the more positive notion of "learned optimism." Seligman was particularly interested in helping people achieve that sense of control and the buoyant sense of well-being and purposefulness that can follow.

Harry wasn't thinking about optimism at all. Quite the opposite. It was the bleaker aspects of learned helplessness that interested him because they seemed to lead toward his goal of true depression. And so he tried another approach. "Again, this was on the inspiration of Bowlby," Suomi says. "Bowlby had described the effect of separating the infant and the mother, the protest and the despair. Back in 1962, Harry had replicated that and Robert Hinde had as well. There was a flurry of mother-infant separation studies. They found the monkeys responded pretty much like Bowlby described but not as severely. The effects were transient. And then Harry came back from Mayo and he had an idea for a chamber that he thought might be useful." Technically, Harry called his design a "vertical chamber apparatus." It was shaped like a narrow inverted pyramid, wider at the top and slanting downward to a point. The monkey was placed in the point, at the bottom of those steep, slippery sides. The wide opening was covered with a mesh. The apparatus worked, as they say, perfectly. The monkeys would spend the first day or two trying to escape, scrambling up the steep sides so that they could look out. This took a lot of energy, though, with the constant sliding and slipping to gain a brief glimpse of the outside. After two or three days, "most subjects typically assume a hunched position in a corner of the bottom of the apparatus. One might presume at this point that they find their situation to be hopeless."

Harry had another name for the vertical chamber; he called it a pit of despair. His colleagues and students tried to persuade him to stay with the technical description. They warned him that it would be politically easier to use less inflammatory, less visual—perhaps less candid—descriptions. "He first wanted to call it a dungeon of despair," says Sackett. "Can you imagine the reaction to that?"

It didn't really matter what you called the apparatus because what really mattered was how it worked—which turned out to be terrifyingly well. You could take a perfectly happy monkey, drop it into the chamber, and bring out a perfectly hopeless animal within half a week. As part of his doctoral dissertation, Steve Suomi ran some of the vertical-chamber tests. As Suomi wrote, in 1970, the chamber changed every monkey who went into it for the worse. It could make abnormal monkeys pathological, make normal monkeys abnormal. The researchers couldn't find even one macaque who seemed to have any defenses against it. Indeed, the pit was a powerful reminder that even a healthy normal childhood doesn't protect against the effects of depression.

In a sense, what the vertical chamber showed, instead, was how naturally we become dependent on the society of others. We live by our intake of oxygen, food, water, and companionship. The monkeys who went into the pit had grown up accustomed to company. "The chamber involved breaking a period of socialization," Suomi explains. Most of the chambered monkeys were at least three months old. They were kept in the vertical chamber for maybe a month, no more than six weeks. The whole point was to take animals who had an established bond—and then break it.

In total, less than a dozen monkeys went into the pit. Two of the animals came from Peggy's nuclear family project. They fared no better than any of the others. When they returned to the lively, friendly hubbub of the family neighborhood, they seemed unable to reconnect. They were withdrawn, slow to respond to others. "Before separation, they had been among the most socially active and dominant of the nuclear family offspring," Harry wrote. Now they were quiet little loners. The monkeys looked—at last—like an undeniable animal model of depression. They looked like animals lost in that hell of loneliness Harry had been working so hard to re-create.

"His work on depression was like a personal metaphor," says Charles Snowdon, then a fairly junior member of the faculty and now head of the Wisconsin psychology department. "He was very depressed in the days of Margaret's cancer. I was brought on as an examiner with Steve Suomi's dissertation and they were using the vertical chambers." Snowdon was appalled by the design of the chambers. "I asked Steve why, why were they using these? And Harry spoke up. He said, 'Because that's how it feels when you're depressed.'"

◦ ◦ ◦

Once they had a model of depression, of course, the charge was to repair the damage. The primate researchers began working with a university psychiatrist, William McKinney. "I basically started my research career in Harry's lab," says McKinney, now director of Northwestern University's Asher Center for Study and Treatment of Depressive Disorders. With McKinney's help, they began probing for the biochemistry behind the disorder.

In one early test, McKinney dosed the monkeys with reserpine, a compound that suppresses serotonin in the brain. Today, of course, we know that one way to treat depression is to boost serotonin levels, keep them elevated in the brain, and some of the best-known modern antidepressants and antianxiety drugs—Prozac, Zoloft, Paxil—employ that approach. But first, researchers had to figure out that serotonin had an influence on depression. The tests at Wisconsin belonged to the discovery period. Monkeys taking reserpine suddenly began to huddle, and their heads drooped as the serotonin levels fell. They were a living demonstration of the neurotransmitter's potency. Who wouldn't watch them and wonder how directly the brain chemistry of serotonin affected mood and whether it could be manipulated?

That's not to say that the scientists in the Wisconsin primate lab made the Prozac connection. They were still trying to figure out whether they even had the right chemistry and what it meant. Harry and his colleagues continued treating depression in monkeys with then-current approaches, testing medications. The researchers found that the existing therapies had definite limits, couldn't break fully through that shell of apathetic misery. The monkeys were, maybe, a little more active, but still withdrawn. They still seemed separate from companions and family. The lab could induce depression, all right, but its scientists seemed a long way from repairing their destructive handiwork.

They were beginning to wonder, though, whether there might be a kind of social feedback loop to depression. You could induce it by ruthlessly removing social contact. Could you then alleviate it also by social means? Perhaps the antidote to taking love away was simply giving it back. One of the most guiding principles in Harry's laboratory was that there was no justification for damaging an animal unless part of the test was to learn how to fix the problem. If one relationship damaged you, could others repair the injury? The Wisconsin laboratory had been working to answer that question for years, ever since cloth mom had proved to be so dismal at raising her charges, ever since Bowlby had pointed out to Harry that loneliness can be next to craziness.

It was Harry's graduate student, Leonard Rosenblum, who devised one of the more compelling tests of the healing powers of friendship. He hauled cloth mom back into the surrogate business and had her raise another four little monkeys. But Rosenblum allowed his infants into a larger circle. Although their home was with cloth mom, for thirty minutes a day, five days a week, they had a play date. His little monkeys rapidly became friends; they were *thrilled* to see each other. When they grew up, they looked nothing like the earlier offspring of cloth mom. They were socially adept, even what you might call normal—outgoing, socially skilled, and group-savvy. Rosenblum compared those surrogate-raised monkeys with youngsters brought up by living mothers. The second group also had play-date time. "What was surprising to everyone was that there wasn't much difference between the two groups," Suomi says. "Every monkey raised by surrogate sucked its thumb. But they could play and get along. When you added in the time with playmates, they became relatively normal monkeys. They had normal patterns of play, they were pretty good parents, they were functional." Other studies showed that if you extended the playtime, you increased the positive effects. If developing monkeys had some chance at normal relationships, they could overcome some of the deficits of life with cloth mom. The healing effects of friendship only emphasized, by contrast, the desperate position of the isolated monkeys.

"The isolates were horribly deficient," Suomi says. "And it was very hard to reverse that." Their next idea occurred at the end of lunch one day, he recalls, yet another session when Harry and his grad students were drinking coffee and tossing out ideas. The isolate monkeys needed a lot of contact to make the turn back to normal. It needed to be gentle contact, steady, soft, friendly. The isolates' normal peers tended to attack these oddball monkeys and then ignore them. But what about really little monkeys, who were almost compulsive clingers, who would adoringly cuddle even with bug-faced cloth mother? Maybe they just hadn't tried the right monkeys. So they matched the isolates with three-month-old youngsters, the most determined cuddlers on the face of the earth. These were the same age as those little monkeys who tried to woo brass-spike mom, who peered lovingly through the Butler box window at their cloth mother. Suomi thought there was another advantage to these baby "peer therapists." They were just starting to become interested in play; they might be able to engage the isolated monkeys in that as well.

To start, the scientists put the baby therapists and the isolated monkeys together for two hours a day. It was almost like watching a peculiar game of tag. The little animals would approach, the older isolates would back nervously away. Again and again, until the unnerved isolates huddled into a corner, heads down, rocking. And then the little therapists would cuddle against them, clinging and stroking. They would repeat this dance until the isolates began to lose their sense of being threatened and became interested instead. Until, slowly, they began responding, just plain old monkey to monkey.

It was Suomi who worked out most of this program for the six-month isolates. With "therapy," the majority could be coaxed back to a functional life. "The only individuals to suffer prolonged distress from these experimental efforts were the experimenters," Harry wrote, in a rare, tacit acknowledgment of how hard it could be to watch a monkey struggle toward a normal social life. But the longer the monkeys were isolated, the harder it was to bring them back. One of the bitterest—and most important—lessons of the isolation experiments is that social skills rust when not used. "Six months of isolation was right on the critical edge of recovery," Suomi says. If the researchers went to a year of isolation, the animals seemed almost warped beyond repair, twisted into creatures that were no longer really rhesus macaques. One baby animal fainted the first time a scientist held him—the sense of warm, living touch was so alien and so terrifying.

Harry thought they might just have to write off the long-term isolates. He didn't like it though. He'd never written off a living monkey; he'd spent his career hoarding them. Finally, one of his newest grad students, Melinda Novak, made a proposal. Novak had also joined the Wisconsin group in the late 1960s, the time Harry's depression was deepening. After graduation, she went on to the University of Massachusetts in Amherst, where she eventually became head of the psychology department. Harry always called her one of his smartest students; so bright, he once said, that she brought tears to his eyes. He was more than willing to listen to her idea. He was more than willing to be wrong if the strange, lost isolates could be saved.

Novak's scheme, moving inch by cautious inch, was a glacial-speed approach to the peer therapy program. She thought the extreme isolates couldn't handle even a couple of hours with a baby monkey. So the long-term isolates were permitted just to see other monkeys, watch them through the bars of their cage. First they watched other isolates, comfortably similar even in their self-clasping and rocking. Then they watched the therapists, who peered back interestedly from an adjacent cage. Then each was given a few minutes a day with a friendly little monkey, then a few minutes more until they were caged full time with a younger, socially competent therapist. It took months, sometimes half a year, before the therapist might coax normal responses from her companion. But under this feather-light program, the young macaques tentatively began to accept other animals.

And eventually, surrounded by friends and family, they began to act like normal animals. You couldn't pick them out of a monkey crowd unless they were suddenly stressed, or briefly placed back in a solo cage, where the shadows of loneliness hovered again. Then temporarily, they would fall back into their old self-rocking habit.

"Melinda's study was remarkable because we believed that one-year isolates were beyond redemption," Jim Sackett says. "The fact that they also responded well to young infant therapists was really a major finding for theories about impoverished rearing."

He and Novak, and Harry, too, thought that some of the techniques developed in the lab, such as peer therapy, might be helpful to people trying to help severely neglected and depressed children. Novak puts it this way: "We learned a lot from those animals—that certain kinds of behaviors could be rehabilitated, that some animals do better than others. Given that kids are reared in so many different ways, so many in a deprived situation, you need to ask those kinds of questions—how robust is the developing system? Is it buffered from certain kinds of experience?"

The monkeys dropped into the vertical chamber were different. They knew how to function socially before they were locked away. It wasn't that they had to be taught how to interact with other monkeys again. They had to want to do that. They had come back from despair and depression and rediscovered the ability—or the desire—to belong. For some of the monkeys it was harder to overcome depression than to acquire social skills. For others, it took only a few days back into the world of companionship to regain balance.

And although it would take them years to follow through the lessons of the isolation studies—from the brief separations to the full year to the depths of the pit—they began to appreciate one of those lovely, common sense results: Everyone is different. When we discuss trauma or grief or isolation, we need to remember that we cope as individuals according to who we are, and that includes our internal strengths and our external safety net. "Perhaps the most important lesson was that not everyone was terribly affected by these experiments," Suomi says. "It took us a while to see it, but quite frankly the vertical chamber experiments led us to recognize that individual variation matters, that it's not just background noise." Novak also finished the isolate therapy with a strong respect for the individual. "The work let us see how flexible the system might be. We know now—better than we did then—that some animals and some people are going to handle these stresses better."

Novak remembers Harry himself, caught in his own depression and moving more and more slowly, like a man in the last stages of exhaustion. "It [the depression] was definitely there and he was tired with it. But even in the cloud of depression, he was quick-witted and he was sharp. You might think he wasn't paying attention. He'd be resting, his head on his desk, and then suddenly he'd raise his head and he'd make the critical point."

Still, there were also days when he simply put his head on the desk and gave in to the same kind of paralysis that numbed his pit-raised monkeys. There were days when the depression was a physical weight that rested on his chest, and it took every bit of his energy just to move it off and sit up. His first graduate student, Abe Maslow, died unexpectedly of a heart attack in June 1970; it wasn't until October of that year, five months later, that Harry finally sent his condolences. "I was saddened to hear of the sudden, untimely death of Abe, just at the time that he had reached the peak of his scholastic career," he wrote to Maslow's widow, Bertha. "I regret my long delay in replying to your letter but my wife, Peggy, has been seriously ill for almost a year, and I have had some problems."

Peggy was definitely getting sicker. "Harry was really tied up," Novak says. "She was the iron horse." Everyone saw Harry as the more vulnerable of the two. Students pitched in to drive him back and forth from the lab, cook his meals, watch and worry about how much he was drinking. Bob Zimmermann was shocked when he met Harry at research conferences in the late 1960s. "I don't think he was ever completely sober around then." For the first time, Helen LeRoy saw the drinking taking a visible toll. She recalls watching as Harry got off an airplane and staggered just a little as he came down the ramp; she wondered then how to help him through this time. LeRoy, Jim Sackett, and Steve Suomi were all working together to keep the labs functioning smoothly. Harry was conserving his energy for his research and for just holding himself together. But Peggy never allowed herself to appear vulnerable the way Harry did. Everyone remembers her as unstoppable to the end. "She was a great model for handling death," Melinda Novak says.

Peggy was uncomplaining, unapproachable. When she had to go to the hospital, she would sit in bed, IV lines hooked up, doggedly editing her manuscripts. When she was discharged, she was back fussing over the nuclear family apparatus. "Everyone talked about how brave she was. She was very ill, she was under chemotherapy, and she would still go up in this attic to check the monkeys, even when she was basically crawling to get there," says Harry's old friend, UC-Berkeley child psychologist Dottie Eichorn.

Peggy didn't want, didn't ask for sympathy—and, perhaps, she found it difficult to find it for others, even Harry's long-time collaborators. In 1970, Jim Sackett was invited to write a chapter on the isolate-raised monkeys for an anthology. "So I went to Harry and I showed him the letter inviting me to do this. I said, 'Really you should do this, this is mostly your work and I just have some things that I contributed, but at the very least, we should be co-authors.' And he said, no, no, you go ahead and do it. I asked him three or four times but I never thought about getting it in writing. So I wrote the chapter. It was a pretty good review of isolate rearing and some surrogate stuff. And then it came out and Margaret Harlow formally charged me with plagiarism."

Sackett was shocked; really, is still shocked. He hadn't worked directly with Peggy and didn't know her well. There were rumors that she resented the amount of attention that he'd been getting. But everyone knew that he'd been standing by Harry for years and, he hoped, everyone knew that he didn't steal other people's work. Others at the psychology department had been hinting that when Harry retired, Sackett might be lab director. To come in and find this notice of formal charges on his desk, well, it stopped his breath for a moment. "So I picked it up and I took it to Harry and I said, 'You remember when we talked about this a number of times and you insisted that you were not going to be involved in this, even when I begged you to?' And he mumbled something and just put his head down on the desk. And that, you know, was a Harry mannerism when he'd decided he'd had enough."

Sackett was forced to go elsewhere for help. He took Peggy's complaint to the head of the psychology department, Wulf Brogden, who dismissed it without question. But winning that round didn't take away the sense of betrayal. It wasn't Peggy's actions that stung so much. It was Harry's. "Harry didn't back me up. He didn't say yes and he didn't say no." At a time when Peggy was dying, Harry had clearly chosen his loyalties. He had no choice, he felt, but to stand by his wife. "It was obvious that was her business and he wasn't going to touch it." Sackett understood the choice. But even understanding the personal nature of it—and so many of Harry's choices were personal—Sackett still couldn't accept that it was the right choice. He started looking for another job and took the one in Seattle when it was offered.

"And, even then, Harry never said a word to me about what had happened. I lost a lot of respect for him over that and I'd had an enormous amount of respect. He was a brilliant, incredibly hardworking man. He had a lot of gifts. But I just couldn't stay."

Peggy was angry at the end. She was never going to finish her exploration of families and children and the way they care for each other. She wasn't even going to see her nuclear family project through. Her daughter, Pamela Harlow, believes her mother had always thought that she would be able to make up lost time. Peggy hadn't grudged the time with Pamela and Jonathan, but she had hoped to rebuild her career. And now that, too, was being taken away from her. "She sat all that time, with all that talent, in the margins of the university," says Lorna Smith Benjamin. "Did she have unrealized potential? There's an understatement. She was an amazing observer, smarter than smart. And she knew it, too—she was angry. It was all-out wrong what happened to her—and there's no other side to that issue."

Margaret Kuenne Harlow died on August 11, 1971, at the age of fifty-two. She had just been made a professor of educational psychology by the University of Wisconsin, some twenty years after her first job had been taken away. Only once did Harry let any bitterness about this seep through publicly and that was during an interview for *Psychology Today*, when he let it be known that Peggy "was not listed as a member of the psychology department until the last departmental budget presented after her death. They thought that made the percentage of women look better."

An interviewer once asked Harry whether Peggy ever tried to compete with him. "No," he said, "there was no competition at all. She knew that I was better at creating research and that she was better at presenting it." In this, though, perhaps Harry didn't give her enough credit. She was methodical in her work and careful in her presentation. "She was very proud of the nuclear family apparatus design," says Gerry Ruppenthal. "And that was the first paper she wanted to do, just describing the device." Her approach was more methodical, less dramatic—but she also had a message worth sharing.

During Peggy's fatal illness and Harry's depression, the Harlows managed to illuminate a near perfect arc of social behavior. In those years at the Wisconsin lab, you could contrast a life rich in relationships to a life having none; compare those sure-footed, confident members of Peggy's nuclear family world to the huddled creatures from Harry's well of despair. You could see the ways that fathers mattered, as well as mothers, siblings, neighbors, friends. You could see how the very biology that makes us rejoice in company makes us, sometimes terrifyingly, vulnerable to losing it.

Harry and Peggy Harlow's studies juxtapose the ways that love can support us and the ways that it cannot. After Harry's depression experiments were finished, Steve Suomi had the vertical chambers disassembled and thrown away. An isolate monkey, he says, will tear your heart out. The chambers have never been rebuilt. The work from that time, though, stands as testament to the ways that love can be the best—and the worst—part of our lives. Harry himself understood that lesson perfectly. It wasn't long after Peggy's death that he began to consider the perils of his position. He knew, all too well, that the cold lands of loneliness are not a safe place to live, not for long, anyway.

NINE

Cold Hearts and Warm Shoulders

If monkeys have taught us anything it's that you've got to learn how to love before you learn how to live.

Harry F. Harlow,
***This Week,* March 3, 1961**

IT WAS AT THIS MOMENT, when he was still stumbling for balance, that Harry Harlow was suddenly accused of being a scientist on the wrong side of truth. It wasn't—as you might think—the monkey isolation experiments that got him into trouble. That would come later. At this moment, in the shifting culture of the 1970s, it was mother love that was the real problem. His pro-parenting stance had turned him into a politically incorrect scientist. He was unprepared, dumbfounded by that turn around. His simplest and most admired work was suddenly on the line. He couldn't, at first, understand it. Love, beauty, truth, motherhood—how could anyone object to that kind of message?

If you were a cynic, of course, and you considered those proclaiming the merits of mother love, you might wonder about their sincerity. The scientific standard bearers were all men. They were all scientists who spent more time at work than at home. They, none of them, had practiced the stay-home-and-nurture behavior that they were urging on women. John Bowlby admitted that his wife took primary responsibility for raising their four children. And Harry had never convinced even his children that they were first in his life.

Beyond the personal behavior standard, there was an edgy, accusatory undercurrent to some of the mother-be-good scientific pronouncements. If you wanted an example—at the extremes—you might consider the stance of Bruno Bettelheim, once a famous child psychologist and now, perhaps, an infamous one.

One of the leading experts in autism in the 1960s, Bettelheim seemed to thrive on challenging others. He rightly campaigned for the better treatment of autistic children. Bettelheim insisted that the children needed individual therapy rather than being locked away in institutions. He took an equally strong position on why those children had become autistic, why they had so much difficulty with life. Autism, Bettelheim proposed, was the fault of the mother. The disease could be blamed on the cold, rejecting mother in particular. Bettelheim had a term he liked to use for these women; he called them "refrigerator" mothers.

Bettelheim visited the Harlow lab after Harry's influential "Nature of Love" talk. As an autism expert, Bettelheim was struck by the rocking and pacing and self-clasping of the monkeys who had been raised with cloth mom. Their restless turning and hand wringing reminded him immediately of his own autistic patients. But he thought he recognized cloth mom, too, with her fixed face and unresponsive body. She reminded him, he said, of those "cold, rigid, intellectual" mothers who induce autism. He thought that cloth mom's stillness and silence, "fixed in space and emotionally unresponsive, prevented the monkey infant from becoming a real monkey," and that the same might be happening to the children he treated. In his book, *The Empty Fortress,* Bettelheim wrote this of the refrigerator mothers: "Certainly they are not free-moving in their emotions or at least not in relation to their autistic child . . . many of them are nearly as frozen, nearly as rigid when they deal with the child as was Harlow's terrycloth mother."

Harry completely agreed that cloth mom had all the limitations of a statue. He didn't deny that her silent stillness could be responsible for serious emotional and social difficulties. But this, he thought, had absolutely nothing to do with autism. Cloth mom's value was in allowing scientists to explore how relationships might alter normal development. He didn't think autism began with normal development. Behind autism, he said, was more likely a brain disorder, still mysterious, undoubtedly genetic in some way. "Possibly some children are rendered autistic by maternal neglect and insufficiency, but it is even more likely that many mothers are rendered autistic because of the inborn inability of their infants to respond affectionately to the mothers in any semblance of an adequate manner," he wrote in a thoroughly dismissive review of Bettelheim's book.

The newly minted feminist movement wasn't impressed. Women weren't inclined to believe even supportive statements from men who spent their lives at work while recommending that women stay home and raise children. Once the women's movement began to emerge as a political power, this male-delivered message of ideal motherhood was, well, infuriating. It directly countered what the feminist leaders themselves were arguing—that mothers shouldn't be shackled to home. Women needed the freedom to go out and to work and to be someone, someone more than their mothers had been. To have prominent male researchers—who had never sacrificed a nanosecond of their professional lives to child rearing or domesticity—tell women that *now* science wanted them back in the house. Who did they think they were fooling? It was obviously yet another establishment attempt to slap females back into place.

Sarah Blaffer Hrdy, herself a primate researcher, a cultural anthropologist, and, by the way, a working mother, recalls that early in her career, an eminent scientist in her field was asked about her work. He replied that "Sarah ought to devote more time and study and thought to raising a healthy daughter. That way misery won't keep traveling down the generations." Hrdy could logically remind herself that her critic was a working father, constantly away from home himself. She could argue that her own husband, a physician, was also pursuing a career, that this criticism was completely unfair. But no matter how hard she reasoned, the guilt stayed. And, she says, it stung.

Other women responded to such charges with simple anger. Women picketed John Bowlby's appearances and walked out of Harry's lectures. Eventually, books such as *Mother-Infant Bonding: A Scientific Fiction* (published in 1993) appeared, in which attachment was dismissed as undiluted psychobabble. By this reckoning, Bowlby's theory was just another attempt to use the tyranny of mother guilt to stop women from living up to their potential. The author, Diane Eyers, freely parodied the rules that men forced on women: "Thou shalt worry that anyone but yourself that takes care of your children will shame you and damage them." When a national magazine interviewed Harry in 1972, the first question dealt with his politically incorrect work. The interviewer pointed out that he, like John Bowlby, had infuriated the fledgling feminist movement by insisting "that human infants need full-time mothers."

Neither Harry Harlow nor John Bowlby handled the backlash well. Bowlby was irritated enough to be wholly undiplomatic: "This whole business of mothers going to work, it's so bitterly controversial, but I do not think it's a good idea. I mean, women go out to work and make some fiddly little bit of gadgetry which has no particular social value, and children are looked after in indifferent daycare nurseries." Harry also used exaggeration as a weapon. He repeatedly told of an event during which he showed a slide of a baby monkey to college students. The male students studied the baby face with interest; the female students, however, let out a breathless "ooh" of response to such cuteness. Harry firmly described that as a natural maternal response. "I have often said that the best way to be a mother is to be born a woman."

He meant that last comment to be deliberately provocative. He had no patience for the suggestion that the mother-child bond wasn't really important, that it was some fiction of scientific misogyny. He'd never cared, anyway, about fitting in or saying the politic thing. He wasn't going to start pandering now.

It wasn't just being attacked that was so upsetting. Both Bowlby and Harlow had endured years of that from their own colleagues. They could shrug off a few insults. It was the irony of it, the injustice. Why should they be criticized for saying that mothers mattered, that the female of the species was loved, needed, and extraordinarily influential? Why should they be harassed for saying that children mattered? Even Sarah Hrdy has expressed dismay over the feminist movement's apparent resistance to the realities of childhood. "Denial of infant needs runs like an invisible and insidious counter current through publications purporting to correct the 'river of mother-blame' coursing through our society," Hrdy wrote recently, in her beautiful and provocative book, *Mother Nature*.

If you looked objectively at Harlow and Bowlby, neither was actually insisting that women should be housebound slaves. Harry had never taken fathers out of the parenting formula. There's evidence of that even in his APA presidential speech. After dismissing the food-equals-love approach, Harry pointed out that if love begins with just being there, with comforting and holding a baby tight, then "the American male is physically endowed with all the really essential equipment to compete with the American female on equal terms in one essential activity: the rearing of infants."

And if you were still fixed on the connection with feeding, Harry added, men are just as capable as women of holding a bottle. If Harry's point seems as obvious as death and taxes, it's worth repeating; as psychology historian Roger Hock notes that, "this view may be widely accepted today, but when Harlow wrote this, in 1958, it was revolutionary."

After the years with Clara and Peggy, Harry was also wise enough not to dismiss women and their fiddly little jobs. It was hard on the family if the mother worked, he agreed; but he suspected that as long as women went home and really paid attention to their children, the relationship would shine through anyway. Harry Harlow had spent more than a decade looking at the whole glittering arc of the social support system. He said that mothers were important, not that they were the be-all of life. He did believe that early experience and care were crucial. And he did think that women were just naturally good—maybe better than anyone else—at giving a child what was needed.

Bowlby also emphasized the need for a strong, stable, loving caregiver, especially in the first three years. He, too, thought women did this job exceptionally well. But what he rated most highly was stability. What bothered Bowlby about other arrangements was the tendency for children to be shuffled from one indifferent caretaker to another. "Looking after other people's children is very hard work, and you don't get many rewards for it," he said. "I think that the role of the parents has been grossly undervalued." He refused to shift from his fundamental point: A baby needs a reliable, loving someone to make the world right. That's what attachment theory is about, after all.

"What did Bowlby actually say?" asks Hrdy. "He said that primate infants, including humans, are born immobile and vulnerable. This is true. He pointed out that they respond very poorly to being left alone, or otherwise being made to feel insecure, which is also true. Human infants have a nearly insatiable desire to be held and to bask in the sense that they are loved. To this extent, the needs of human infants are enormous and largely non-negotiable."

Hrdy belongs to the modern generation of scientists, but Bowlby could definitely have used her on his side back then. So could Harry. If Harry could have made his argument with Hrdy's clarity, and perhaps her charity, he might yet have won at least some of his female audience back. He didn't try elegant persuasion. He was tired and unhappy. He had lost his wife. He was battling alcoholism and loneliness and the clinging dark grip of depression. He was listening to women say that the best of his work was merely male tyranny, without merit or honesty. Quietly and thoroughly, Harry simply lost his temper. He didn't bother trying to educate his audience again. He turned to the other weapons he possessed—a talent for being provocative, an ability to underline his words with gilt-edged sarcasm. "Harry's pen was sharp," says a former student. "But his tongue was equal."

When a bright young psychologist named Carol Tavris came to interview Harry for *Psychology Today*, she found her subject ready for battle.

"If you don't believe that God created women to be mothers and essentially nothing else, let me prove it to you," he told her. And that was only in the second paragraph of a ten-page interview published in the spring of 1972. Harry mockingly went on to contrast the play style of the male and female monkeys. "Males play rough and females play soft and sweet and gentle. They sit quietly on the sidelines saying mean, catty, nasty things about other women.

"Physical strength is the one trait in which man is superior to woman and speaking is the one trait in which woman is superior to man. Now consider what happens when a couple argues. The man tries to *talk* to the woman. The stupid fool, he can never win. Are you married?"

TAVRIS: Do you have to be married to argue with a man?

HARLOW: God created two species, one named *man* and one named *woman.* I can even tell you the difference between them. Man is the only animal capable of speaking and woman is the only animal incapable of not speaking.

TAVRIS: Women's liberation will get you for that one.

Even today, Tavris isn't sure how much of this was real and how much of it was Harry Harlow's wish to provoke.

"Sometimes Harlow was a blatantly appalling sexist, yet it was hard to know how much of it was designed to rile people up and how much was what he really thought. There was an unexpected sweetness to him that made his obnoxious remarks seem oddly artificial at times," she says now.

Harry wrote her a letter after the interview appeared:

Dear Carol:

Let me congratulate you on the splendid job you did on the interview. I am not sure whether you were interviewing me or I was interviewing you. If I had known as much about recent developments in primate research as you do, I would have been able to respond to your questions in a more intelligent manner. I am convinced that you could raise the IQ of a vegetable—human, or otherwise.

Her editor, after reading Harry's note, sent Tavris a memo: "Total surrender. But not the letter of a small man."

She agreed.

In fact, Harry did hold men and women to the same professional standard. He expected them all to work really, really hard. Harry didn't hit up his female students. He didn't demand sexual favors. He didn't chase them around the cages. He didn't play groping games in return for good grades. He did expect the same mental toughness and independence of the women that he expected of his male students. "He'd push you to your limit if he thought you were up to it," Lorna Smith Benjamin says.

Then Lorna Smith, she came to the lab in 1956, one of the first female graduate students there. She wasn't invited to have a drink at the corner bar. Harry was comfortable doing that with the male students, she says. But he never talked to her "like a girl," demanded less of her, or treated her as if she needed coddling. The paradox, she says, was that Harry treated everyone equally and women just weren't used to it. "His manner was challenging and hostile and females of the time were not used to dealing with that. If they were flattened and intimidated, Harry would take advantage of that like any good primate. If you fought back, that was fine with him. He was always mocking and sarcastic and he didn't mind getting it back; but women, especially women back then, weren't socialized that way."

Smith once put a drawing of Harry's head and a monkey's head in his office, with a "Paradise Lost" slogan over them. Her friends in the lab warned her that he would be angry. Instead, he roared with laughter and kept the drawing for his own amusement. "I'm not saying that I was not a victim of sexism and sexual harassment in that time period. I was indeed," Smith adds. "But none of that was from Harry. He never propositioned me or grasped or even flirted, and therefore there was none of that retaliation for refusing." Even now, after years as a respected psychologist at the University of Utah, she appreciates that her graduate professor judged her strictly for her abilities.

"He liked independent women," Melinda Novak says. "Harry said what was on his mind and he expected that from you." He liked to tease, though. She was running the slide projector in class for him one day—because he never could figure out how to use it—when he suddenly said to the class, "See Miss Novak there, I'm going to marry her tomorrow." "I laughed, the class laughed, it was a good way to see if we were all awake." Harry encouraged her research, supported experiments that she had designed, boasted about her to others, and, when she left, he gave her monkeys and cages to give her a head start on her new career. "And I have to tell you, there's no support in my life for the idea that Harry was sexist. People would say to me, 'Wasn't he sexist?' And I'd say, 'Are you crazy? He was terrific to me.'"

"But I will tell you that what he did used to do, what Harry liked to do, was to get a rise out of people. I'm not sure why, but he wanted this dialogue. One topic he liked to push buttons on was religion. He'd be at a religious school and he'd say something controversial. One of Harry's favorite lines, and he used it many times, was to begin a lecture with a picture of two monkeys copulating, one atop the other, and then announce that he was going to call the talk 'The Sermon on the Mount.'"

His mounting joke was a regular feature in his talks. Sometimes people laughed. "He had the most perfect comic timing," recalls McGraw-Hill psychology editor, Jim Bowman, who regularly attended APA conventions. "He'd put up the slide and kind of drawl, 'I call this . . . The Sermon on the Mount.'" And sometimes, as friends and colleagues remember, people stood up and left when Harry told that joke. Gig Levine remembers a psychology meeting in Germany in which the audience simply sat, unsmiling, in disapproving silence until Harry's voice softened almost into a whisper as he finished his standard talk. Mostly, though, Harry couldn't resist those wry sexual innuendos. They would come spontaneously into his head. Even at a session focused on the behavior of worms cut into regenerating segments, Harry's first comment was "I'm glad humans aren't the only animals to lose their heads over a piece of tail." It was such off-hand comments and off-color jokes, Novak says, that began to get Harry into trouble. "People would call up the APA. They'd try to get him kicked out. They'd call me up and ask for the skinny on how he mistreated women. And they just got it wrong. He made controversial talks. Sex was something that Harry would use to get some kind of funny reaction."

Harry believed that the sexes were biologically different in ways that mattered. Once again, he couldn't have been more politically out with the fledgling women's movement. The more popular position was that males and females might look a little different physically but they were basically, fundamentally, alike. If women behaved differently, that was only because a sexist culture had taught them to behave differently. Biologically—barring some sexual apparatus issues—men and women were the same. Logical political sense stood behind that stance and it had to do with arguing for equal treatment.

To Harry, the women's movement was just wrong again. He'd been watching natural sex differences for a good three decades. Occasionally, he'd even made a point of telling people about it. Back in September 1961, during a speech to the APA in New York, he pointed out one such variation. Female monkeys are far more likely to stroke and pet when they groom another monkey. Males are brisk and businesslike, efficiently digging out dirt and bugs. "Caressing is both a property and prerogative of the girls. They show better manners too," Harry said. Little male monkeys were more prone to make faces at each other, especially threat faces: "The females rarely make threat faces and almost never at the little boys they play with."

If he hadn't been aware of those sex differences, some of his most important studies might have failed. When Harry and his students were trying to rescue troubled monkeys, especially the long-term isolates, they used females as peer therapists. The nervous and neurotic isolates needed to be stroked and gently handled. Threat faces and chase games would have only unnerved them further. If at this moment of selecting therapists, Harry had ignored male-female differences—or worse, pretended they didn't exist—he doubted that the isolated monkeys could have been brought back into normal range.

That didn't mean that the monkeys were exact models of human children. But, as he emphasized in his speech, there were very clear parallels. In monkeys, Harry said, "real rough and tumble play is strictly for the boy monkeys." The little females played, too, but left to themselves they chose less physical games such as chase or tag. Sometimes they were provoked into the games by small male monkeys, who seemed to chase the females around the cage for the sheer fun of it. Any parent, he said, could observe a similar pattern in human children just by watching boys and girls play. "There is no fundamental difference between a Madison park and a laboratory monkey playroom." In his speech, he told the story of attending a second-grade picnic and watching almost identical behaviors: little boys screaming round the park, little girls chatting to each other. "No little girl chased a little boy, but some little boys chased little girls."

Harry had no doubt that, like other childhood play patterns, these sex-typed ones could predict adult behavior. The chatty little girls might grow up to be talkative women. The rowdy little males might become competitive and, perhaps, aggressive men. The nurturing behaviors learned by small females in childhood might carry over into the way they behaved as adult mothers—or friends. The adult males might continue to hone their skills at game playing and alliance building. They would probably continue to be less cuddly than the females. That was just the way it was. Biology influenced behavior and vice versa—and to Harry, it wasn't worth arguing over. If the feminist movement flatly refused to consider that connection, well, it only added to his contempt for the movement. He expressed that once, in a pointed bit of doggerel: A woman's libber's not a saint / She's just a girl with a complaint / The sexes aren't created equal / A tragic story with no sequel.

Harry never seemed to realize, entirely, that this wasn't just a matter of silly politics. The issue mattered, much, to many people. It wasn't right that bright, dedicated scientists such as Sarah Blaffer Hrdy should be accused of poor parenting when they were trying harder than many of their male colleagues to balance work and family. Such injustice seemed everywhere and the newly charged female awareness was prickly, defensive, and resentful. Women were poised to detect insult. As psychologist Carol Tavris put it, "Everyone's consciousness was on hyperalert for signs of sexism." And men, if they were reluctant to change what they did, were at least softening what they said about women. They saw danger signs out there. Many men were beginning to tiptoe a little more cautiously through what was clearly a well-mined political landscape.

Harry ignored those signals. He continued saying what he thought, continued telling stories that had worked for him in the past; and he deliberately baited those who didn't understand his approach. It wasn't in him to give up on a good point—or a good joke—just because of an emerging sense of squeamishness about women. If his audience was outraged, well, that "tickled his perverse and sometimes waspish sense of humor," Melinda Novak says.

He didn't hold back. Even Harry's closest colleagues, such as Steve Suomi, worried about his contempt for moderation. Far too often he seemed to cross the line between being provocative and being offensive. Suomi wondered whether Peggy, that vigilant and intelligent editor, had been literally holding Harry back for years. Harry hinted as much to Tavris. His wife had refused to let him use some his best lines, he complained. Now he dusted them off, jokes that he'd been holding back for years. Helen LeRoy, his long-time friend and lab colleague, also tried to tone down his remarks. But they just kept bubbling out.

Harry published an autobiographical history of the surrogate mother that looked backwards to his zoo days with the old orangutans Maggie and Jiggs. The tale included a story about the irritable Maggie when she was taken on soothing walks by the zoo director. One day, Harry said, a child threw a rock at Maggie. The ape, angered, reared up into attack position. Desperate to stop her from injuring the boy, the zoo director looked around and saw the kid's baseball bat on the ground. He picked it up and whacked Maggie with it. The orangutan halted, rubbing her head. She then put her hand back into the zookeeper's and "let him lead her to her cage where she looked at him with loving admiration." At last, Harry wrote, Maggie had found "a man who understood the psychology of females."

In a paper on surrogate mother experiments, "The Nature of Love—Simplified," Harry discussed the importance of temperature in the surrogate mothers. He and his students had designed an extra-warm surrogate and a cooled-down one to check for the baby monkey's responses to mother's body temperature. "We felt we had really simulated the two extremes of womanhood—one with a hot body and no head and one with a cold shoulder and no heart," he wrote. When Harry wrote about the isolated monkeys and their sexual incompetence, he described the restraining device that would harness a female so that a male could mount her. Harry didn't call it a restraining device, though, or a reproductive apparatus, or any of those neutral terms that animal researchers like to use. This was, after all, the man who came up with "pit of despair" to describe depression experiments. So Harry informed his readers that the restraining device was "affectionately termed the rape rack."

He also sprinkled such comments through his speeches. Jim Sackett still shakes his head over Harry's guest appearance at the University of Washington, after Sackett had become a professor there. "Did you ever see *The Producers,* that scene when the audience actually gets that the story is Hitler? That's what they looked like here." Scientists and students alike had their mouths hanging open; they were turning to look at each other in disbelief. "They were aghast. It could have been them in *The Producers.* He put all his sexist remarks into one talk, I think, and all the women's libbers, remember this is the 1970s, they just left."

Jane Glascock, then a Ph.D. student in cognitive psychology at the university, talked excitedly with her friends when they learned that the famous Harry Harlow would be speaking. They were expecting to hear about love, about connection and relationship in its purest, warmest sense. Instead, the message came laced with antifeminist mockery, "derisive, sexist remarks of the most insulting and unsubtle kind." Glascock listed them in a furious letter to her department head: Harry had said that "isolation-reared monkeys were forever confined to a stage of infantilism, which wasn't so bad if you were a female." He had praised Melinda Novak by saying she was so bright that "it broke his heart she wasn't a man." He had talked of hope springing eternal in the human breast, and then showed a picture of a nude woman: "Dr. Harlow obviously has his own problems. However, by playing for laughs by degrading and insulting women for want of substantial research matters to present, he has insulted all of us and made those who sponsored him appear nearly as foolish for doing so."

Glascock was sitting in the audience with Earl Hunt, her major professor, and, she says, "I remember him pretty much physically restraining me during Harlow's talk. This was still fairly early on in the women's movement and I was just so shocked and outraged by his behavior and his allusions. So I ran right home to my typewriter." Looking back now, she's less angry. Harry's talk might have been one of the era's more outrageous instances of sexual offensiveness, but hardly the only one. Harry Harlow, as she rapidly learned, was just one among many men slow in adjusting to the new realities. One film shown to grad students, on the subject of perception, involved the narrator at a party primarily following "this bimbo with these immense breasts while trying to make points about psychology. It was horrible." In her memory, now, Harry is just one of the old boys, men of an earlier generation, who hadn't been able to make the change when the world around them did, who were still learning, with shock, that women should be treated with the respect routinely given to men.

There were moments when you might forget that Harry Harlow was a man who liked and appreciated smart women. He could sound as if women were not only a lesser half of the species but a drag on the better half. California psychologist Steve Glickman still recalls one such acid-edged conversation. Glickman, a professor of psychology at the University of California-Berkeley, is a noted expert on the link between hormones and behavior. After one particularly drunken party following a psychology conference in Chicago in the early 1970s, Glickman and his new wife, Christa, drove Harry home. It was 2:00 A.M. when they got to the hotel. "And Harry's still not ready to have the evening end and he says, 'Let's have breakfast.' And we're sitting in this booth in the hotel coffee shop and all of a sudden Harry leans forward, avoiding Christa's eyes, and says to me, 'Now that you're happily married, you'll never get anything done.'"

And yet Harry was a man whose career had flourished in the companionship of bright, capable women. He'd married two of them, after all. In the time since Peggy's death, there were once again signs that Harry was a man lost without a companion. His friends and colleagues and students, and even his children, pitched in to help. Helen LeRoy and Ken Schiltz, fed up with his shabby clothes, took Harry shopping and outfitted him in style. Harry continued to put in the extra hours at the lab. But he wrote to former students that he was finding life without Peggy extremely difficult. He was fixed on the world of twosomes, happily married or not, around him. He was going to faculty parties again, asking people whether they were married. "After Peggy died, Harry was fascinated by people's relationships," says Wisconsin psychologist Charles Snowdon.

Harry had never liked being alone. He knew the importance of companionship and comfort. He longed for it. He started thinking not about some new, bright relationship but about an old one. He'd regretted failing at his first marriage. He began to think now about starting over. He and his first wife had always stayed loosely in touch because of their two sons. Harry had not been a stay-at-home father but he'd never chosen to disconnect from his older sons. Peggy had always insisted on his keeping a distance from his other family. But when Harry had business trips that brought him nearby, he'd visited his children anyway. His older son, Bob, had continued to write to him: "He always wrote back. The fastest reply was when I called him Harry in a letter. I was twelve. He wrote back and said, 'You call me Dad.'"

At the time when Harry began thinking again of Clara, she was also on her own. Clara had remarried twice. Shortly after the divorce from Harry, she had chosen a very different kind of partner. Robert Potter had studied at a technical school and was employed as an industrial parts salesman. The Potters decided to move to the Southwest. Clara had lived there as a child and loved it. They bought a small ranch outside Reno. Clara had wanted a full partner in parenting and now she had it. Her second husband took his new fatherhood seriously: "He was shrewd, tough, a strict disciplinarian, extremely fair," says Bob Israel, looking back. "He was a wonderful man and he cared a great deal about us. Not the hugging and touching type. But he was always there, and if we wanted an opinion, he would listen and he would say what he thought." To Harry's outrage, Bob and Rick took the last name of Potter. In an angry note to the clerk of the court in Dane County, where his divorce was handled, Harry wrote that he would continue to pay the child support. But, "I make this payment and any subsequent payments under protest since I have learned that the children have been living . . . under assumed names."

The Potters didn't stay in Reno. They had a child there, a little boy named Thomas, who was born in 1949. Two years later, the child drowned in a drainage ditch behind the ranch house. They simply left, fled the place, moved to the Carmel Valley in California. They started a children's clothing store and named it Little Tyke in honor of their dead son. Clara continued answering Terman's questionnaires, but there was little trace of her early joyfulness. After Thomas's death, she mailed back forms full of blank spaces as if she didn't have the energy or heart to write about her life.

She had more trouble to come. Robert Potter suffered from bleeding ulcers. In 1960, after a series of operations to try to mend his stomach walls, he died of a resulting infection. Clara decided again on a new start. Searching for a complete change, she decided to try hotel management school, was accepted into a program in Tennessee, packed up, and moved again. She married a fellow student, Clint Thompson, but the marriage was brief. "He was a nice guy," recalls Bob Israel. "But whenever he touched alcohol he was just out of control." She divorced Thompson in 1965, after two years, and took a job as a counselor in the University of Tennessee's school of nursing in Knoxville. Clara Thompson was working there when her first husband, troubled by regrets and loneliness, came calling.

They picked up the old, good relationship with startling speed. "I've always said that Dad never really stopped loving Mom," Bob Israel says. Harry felt, at least, that Clara knew him, for better and for worse. He wrote her a poem to that effect: "The things that cause you no surprise / are all my lies and alibis / for you can all too easy see / the faults that are a part of me."

They were remarried March 1972, eight months after Peggy's death, in a small civil ceremony in Knoxville. The newlyweds celebrated by partying late into the night with an old psychology friend, University of Tennessee psychology professor William Verplanck. The reunited couple then went on a honeymoon tour of England, Scotland, and Ireland. Harry wrote to Verplanck that "we plan our third honeymoon in Hawaii. We are too old to pass up any chances since the next could easily be our last." He also thanked Verplanck for his company on the wedding night and sent him a bit of Harryesque verse: "Courage strong and honor bright / Courage usually lasts the night."

Clara's reports to Stanford turned suddenly exuberant again; she wrote of her "unexpected happiness in remarriage to my present husband." The Harlows settled into a condominium on a newly developed street of apartments, condos, and small shops in the western suburban edge of Madison.

By this time, though, Harry had been in Madison for more than four decades. The NIH primate center had another director now, named Robert Goy, who had his own plans for the place. Harry admitted to his friends that he was tired. Even the prospect of new monkey experiments suddenly lacked its former appeal. Harry wrote to a long-time friend, Duane Rumbaugh, a Georgia State psychologist, that he would like to compare the abilities of rhesus macaques with gorillas and chimps. "The only reason why we are not doing it is that we are bankrupt, financially, mentally, and emotionally."

There were other reasons to feel weary. Harry was noticing an odd shakiness, the occasional unnerving loss of balance and focus. His doctors would warn him that this looked like the start of Parkinson's disease. They would see what they could do to slow it down with medication.

Meanwhile, Clara was remembering everything she had disliked about Madison's weather. "Normal human beings can't live in this god-awful climate very long," Harry told a local reporter. "Wisconsin has a highly humid environment and many people should not live in a highly humid environment," he went on. "For example, it gives me mild to intolerable rheumatism. It gives my wife hopeless asthma." As the damp air continued to trouble her breathing, Clara pushed harder for a move to a better climate. She still loved the dry, bright air and coppery landscape of the Southwest and preferred the region to any other place she had ever lived. She urged Harry at least to look there. He wrote to some of his former students: John Gluck, now at the University of New Mexico in Albuquerque; and two others, Jim King and Dennis Clark, who had settled at the University of Arizona in Tucson.

All three responded with open invitations. He and Clara decided to take a vacation exploring the Southwest—and its universities. As it turned out, the psychology department in Arizona had come up with an ingenious way of recruiting well-known faculty at almost no cost. It would offer unpaid "research professor" positions to retiring scientists who had achieved great acclaim in their fields. "You know what happens to giants," Harry joked. "They go to seed." These well-known researchers could have an office, work on what they liked, bask in the clear Arizona light. They could also add to the prestige of the Arizona university with very little cost to said university. Neil Bartlett, who was then the head of the psychology department, recalls urging Harry to choose his school over New Mexico. "And I said, well come on to Arizona because you have two students here," Bartlett recalls. "There's only one in Albuquerque."

Meanwhile, to Gluck's exasperation, his department head hesitated to hire a faculty member more famous than he was. And so Jim King suddenly heard from his old professor: "Harry just called me up one day and said he was coming." Bartlett had not firmed up the position but, after King called him, he hurried over to the administration offices and settled it. "I told the president that Harry would bring recognition to the university."

In 1974, Harry resigned from the University of Wisconsin. He insisted that this was mainly about the medical reasons. The Harlows were "condemned [by their health] to live in Arizona," he explained to a local newspaper. Truthfully, though, Harry would rather have stayed. He had spent most of his life in Madison—it was home. He might joke about the weather but here was the lab he had worked so hard to build, his good group of graduate students, his closest friends and long-time colleagues. All his best work had been done here, memorable results accomplished against long odds. It hurt him to leave; it was like the physical wrench of walking away from a love affair. "I have an enormous affection for Wisconsin," he admitted. "You can't teach in a school for forty-four years and not have some affection. You can't be married for forty-four years and not have some affection. It's the same thing." Leaving Wisconsin was like leaving himself—it was here, after all, that he'd reinvented love and, really, invented Harry Harlow.

Of course, he started adding to the lore as soon as he moved. Byron Jones, now a professor of biobehavioral health at Penn State, was a graduate student at Arizona then. He still remembers the day Harry arrived. Jones was standing in the department office when the phone rang. The secretary turned to him and said, "There's some crazy guy on the phone that wants someone to pick him up at the airport."

Jones asked her who was calling.

"Harry Harlow," she replied.

Jones grinned. "Tell him we'll be right out," he said.

Harry and Clara bought a condo in a place as unlike Madison as possible. Their new home was part of a Spanish-style development; the homes had white stucco walls and red tile roofs and were tucked into a planned landscape of graceful palm trees and magenta-brilliant bougainvillea. It didn't take Clara long to begin dressing up their life further. She bought herself a fur coat. She dressed Harry in tweedy sports jackets and elegant suits. She'd collected crystal, more than enough for all occasions, during their trip to Ireland. "They had two china cabinets filled with Waterford," says Penny King, an Arizona schoolteacher and Jim King's wife. "And Harry was natty. She kept him a fashion plate." Clara ruthlessly restricted Harry's drinking. She insisted that he exercise. Harry routinely rode a city bus part of the way to the university and walked the remaining mile and a half. She'd drive to campus later and work at the library and then pick him up in the evening. "He was in great condition, especially compared to what he'd been like at Wisconsin," the Kings agree, almost in a chorus.

Harry reintroduced himself to his sons. Bob, by that time, had changed his name from Potter back to his father's old family name of Israel. Bob Israel was working as a fundamentalist preacher, near Portland, Oregon, and he was surprised and delighted to find that his father would support him in that calling. "I think this is where your heart is," he still remembers his father saying. "When I saw him again it felt like we hadn't missed a beat." Rick Potter lived closer, in nearby Phoenix, where he worked for the state government. But although he saw his father more often, he didn't have that sense of picking up an old relationship. The years apart were too long and the memories too few and too scattered. "He could be disarming," Rick Potter says. "But we never were father and son. Over the years in Arizona, we were two grownups." They got along fine, he says. But

Rick never forgot which parent had been there for him. "My mom was the one who loved me and spent time building that bond."

Clara had had many years to think about the collapse of her first marriage with Harry. And she believed that Harry's love of psychology meant that he couldn't maintain a relationship with someone outside its charmed circle. She was now convinced that it was the University of Wisconsin's refusal to let her continue with psychology and the "change of vocation that had led to the divorce." Clara didn't plan to make that mistake again. She didn't want to. She had dreamed of being a psychologist. Now she had a chance to win some of that dream back. As Clara Harlow, she had been given the title of research associate at Arizona and a carrel in the library for her work. She had obtained recommendations not only from Harry but also from Stanford University psychologist Robert Sears, who had become the keeper of the Terman gifted study.

Sears wrote a warm letter on Clara's behalf. He had known Clara since she was a graduate student and, he wrote, "in the early years she was one of the brightest and sharpest young women I knew in the psychology area." Clara told Sears that she hoped eventually to be recognized on her own, to see "if I am approved without being under the shadow of the name of the master."

Clara had an idea of her own about childhood play. She wasn't thinking of it in the way Harry and Peggy had, as an exercise in adult behavior, or as a way of negotiating and building friendships. She was thinking more of the mechanics of motion and what they accomplished. In a sense, her concept follows the moving surrogate mother idea that Bill Mason had explored, that one needs physical motion for healthy development. Both the Harlows admired that work. From Arizona, Harry wrote to Bill Mason, telling him that he was perhaps the smartest "surrogate graduate student" that he had ever worked with. In the same letter, Harry added that he thought Clara showed that same kind of promise.

Clara was interested in the times that we just play by ourselves. After all, if no friends are available, we may skip and dance, tumble and swing all on our own—and all to the good. Perhaps "self-motion" play, as she called it, is also part of building that strong body and capable nervous system. At least that was a primary argument in a paper that Clara wrote with Harry as co-author. The paper was published in the Proceedings of the National Academy of Sciences. And, if the idea that dancing alone may sometimes be biologically necessary came from Clara, the wry description of how it works came distinctively from Harry: "Human self-motion play takes place primarily outdoors. When it takes place indoors, parents protest."

The article was a beginning. But, even more, Clara wanted to work with Harry on a book, a definitive book. People had been trying to talk Harry into writing definitive books for years. During the early 1960s, Harry had been the psychology consultant for McGraw-Hill's psychology tests. His old editor at McGraw-Hill, Jim Bowman, had coaxed and teased him to write a major book on his work, rather than just polish the contributions of others. Bowman thought, still thinks, that Harry Harlow was one of the smartest psychologists he ever knew. "He and I talked so many times. He was going to do a book. He was going to do a lot of books. It's really interesting to me that he never did a big book. Because he could have."

Even in retirement, Harry kept receiving the book proposals. Bowman, retired from McGraw-Hill, still urged one more try. At his request, one of the younger editors in McGraw-Hill's textbook division, Tom Quinn, stopped in Tucson to do some recruiting. "And Harry said, sure, he was interested, but I didn't think he was really serious," Quinn said. "He talked about some ideas he had but I had a hunch it wasn't going to happen. I sent him a letter and I don't recall ever hearing back." Another of Harry's former students, Stephen Bernstein, now at the University of Colorado, also thought he should do a book. The book Bernstein had in mind was actually the kind of book Clara was suggesting, a collection of Harry's major works—the surrogate studies, the learning sets, sex differences, depression, the whole panorama of love. At one point, Clara suggested an almost lyrical title for that volume, *The Lands of Love,* but Bernstein had a more pragmatic project in mind.

When Harry and Bernstein reconnected at a meeting in Switzerland, Bernstein remembered all over again what a "W. C. Fields kind of character" Harry Harlow was—they'd be driving through some quaint village and Harry would roll down the car window and shout greetings to startled pedestrians and farm animals alike. He was also reminded of what a good scientist his former professor was. He was afraid, without the book, that people wouldn't remember that. There needed to be a record, Bernstein insisted, a place where people could find Harlow's collected work. Harry agreed to Bernstein's idea as long as the editing included Clara. Bernstein agreed.

As the project grew, however, Bernstein sometimes regretted that commitment. He didn't find Clara nearly as charming as he found Harry. He found her defensive and possessive. "We didn't get along well," he says simply. Back at Wisconsin, both Helen LeRoy and Steve Suomi also had struggled to adjust to Clara's new role in Harry's life. She was warmer and friendlier than Peggy had been, but, in her own way, equally as tough-minded. Suomi and LeRoy had both worked consistently with Harry, even during his marriage to Peggy, reading his drafts and helping improve them. Peggy had been happy to have their input. Both she and Harry thought that Helen, in particular, was an outstanding editor. But Clara didn't take their suggestions as helpful. She took them as criticism. Shortly after she began working with Harry, Clara asked him to let her handle their manuscripts herself, without being second-guessed.

"She thought she could be another Peggy," says Suomi. "But she wasn't, not in terms of academic training or of knowledge. Some of us were uncomfortable with her co-presenting with Harry, without the requisite academic credentials. A psychoanalyst might say this was her way of competing with and surpassing Peggy in Harlow's eyes and with the rest of the world." LeRoy, too, thought Clara was capitalizing on Harry's fame, but "I'm not sure that it is fair to say that her later attempts at being a scientist were to compete with Peggy, but rather were Clara's efforts to prove herself to others."

Bernstein acknowledges that the resulting book of Harry's collected works, *The Human Model,* wouldn't have been finished without Harry's wife as a collaborator. "The book does owe to Clara," he says. "Harry was declining physically by then. She made it happen." Perhaps it was a good thing that Harry hadn't taken on a bigger book project because he was now, in the late 1970s, starting to get sicker. The drinking, the smoking, the short nights and long lab hours, the depression, Peggy's death, leaving behind his life at Wisconsin, the Parkinson's disease, it was all catching up with him at once. He suddenly, almost abruptly, slowed to a stop. "He didn't even do much writing," says Dennis Clark, his former student then on the Arizona faculty and now a Tucson businessman. "He'd go in the office where they had a drip coffeemaker and he'd pull out the carafe before it was full and let coffee run all over. We often wondered if the drinking had done something to him." Harry wondered that himself. He had cut back, at Clara's insistence, but he couldn't help suspecting that years of drinking had worn him out. "You would think his liver would have totally shriveled," said Jim King. "There was all this alcohol he consumed, the almost continuous drinking."

Harry was so worried about that himself that he went to the doctor to have liver tests done. After the results came in, he told King about them, almost in disbelief: "You know, my liver is totally normal. I can't believe it," he said. King still can't quite believe it himself, that Harry treated his body so badly and held together as long as he did. "I think Harry had good genes—in terms of his liver and in terms of longevity."

It was really the Parkinson's disease that was steamrolling right over everything else. Harry was still taking medication for the disease but he was no longer responding to the drugs very well. He didn't talk about the disease much, but when he visited his old colleagues they were shocked. He traveled to Tennessee to give a speech to the psychology department, urged by Clara, and he could hardly stand upright without help. His old friend, Verplanck, was shocked by how fast he seemed to be aging. During his notorious talk at the University of Washington, the psychology department head, Earl Hunt, had thought that the old Harry was flickering away, that sitting through the speech was like watching "an unfortunate act of a sick man at the end of a distinguished career. It had nothing to do with the great work he had done." Harry gave a speech at Stanford and dismayed his old colleagues by blistering his hand while trying to hold a match. "When he was here, his hands were shaking," Stanford psychologist Eleanor Maccoby says. "He was trying to light a cigarette and he held his finger in the flame and burned it. I don't think he even noticed. It was very sad."

"As it got worse, he barely talked at all," King says. "The shaking was better but he had the mask-like face of later Parkinson's and he got quieter and quieter." He would still come to work, but there was no more talk of studying monkeys or of intelligence tests or of yet someday writing the big book of love. Harry would retreat into his office, writing brief answers to letters, dreaming up new rhymes. "He was lonely," King says. "He had this office but there wasn't a great deal he could do. He sat in the office, he read things; he wrote doggerel."

We have to think of sour-faced-Dan
Who, being a Parkinsonian,
Could never laugh to show his glee,
Or work upon one's sympathy
By looking sad when feeling pain.
(He often tried but all in vain.)

There were still flashes of the old Harry. "You know there was a head secretary there and he'd give her the verse to have the secretaries type it and she was just outraged by this. Somehow Harry heard about this, and so he would give the doggerel directly to the other secretaries. And you know, it was much more interesting than what they were typing ordinarily and he'd bring them bottles of liquor as a thank you, and I think they enjoyed it."

The only reason birds can fly
Is they have faith and dare to try
Of course, they're helped by subtle things
A fan-shaped tail and pair of wings

In the summer of 1981, Harry's memory began to fail. He was confused, and even hallucinated. He was in hospital for ten days in July and August and another six days in September. When he came home he was still bewildered. Once when he stumbled into the home of an elderly neighbor, the frightened woman alerted the complex with her screams. Clara decided to put him into a home. "The last few months have been a nightmare with Harry in a nursing home," Clara wrote to Bob Sears, director of the Stanford gifted study. In November, she said, the doctors thought that Harry had suffered a stroke because he went into a coma and never surfaced again.

I once approached the pearly gate
And wanted in but was too late

Harry Harlow died on December 6, 1981, at the age of seventy-six. After his death, Clara edited another collection of his works, part of a centennial series on high points in psychology, that was put together by the academic house Praeger. She wove into it Harry's research from beginning to end and she worked on it with complete determination. She wrote again to Sears and told him that her eyes were failing, they ached and blurred, but she was absolutely going to finish the book for Harry. It was published five years after his death under the title *From Learning to Love,* and Clara began it this way: "Harry Harlow was not always famous but he was always unique."

TEN

Love Lessons

Love, and the lack of it, changes the young brain forever.

Thomas Lewis,
Fari Amini, and Richard Lannon,
***A General Theory of Love,* 2000**

A MOTHER'S FACE IS ALWAYS beautiful. Harry Harlow came to believe that, years ago. He couldn't design a mother's face that would turn a baby away, not even cloth mom with her red stare and her flat green smile. In the look of her mother, the infant saw the gorgeous appearance of security, the commitment of just being there. No scientist has ever found an object in the universe that a baby would rather see than a mother's smiling face. Perhaps there's a carryover effect; beyond mom, babies love to look at faces, period. Since Harry's first mother-baby work, since Bowlby's theory was accepted, since *The Competent Infant* and the scores of other books exploring child development, psychologists had come to marvel at how passionate babies were about nature's assembly of eyes, nose, and mouth.

Scientists can, and do, show these very small humans pictures of trucks, trees, animals, flowers, people from head to foot—and infants will look at them all. They're interested. They'll study the scene, the colors and patterns. Then they'll turn back to gaze at a pictured face all over again. Curve of lips, arch of brow, narrowing of eyes—there are countless meanings in this human canvas. A baby will peer intently and try to decipher those flickering expressions. In systematic tests where infants are shown pictures of people with varying expressions, researchers find direct evidence that the infants deftly interpret facial meaning. The babies prefer joyful faces to angry ones. And they respond. Very young humans stare happily at a beaming smile, look somberly back at a frown.

Babies scan faces, it seems, for answers to their most important questions: Am I doing the right thing? Am I making you happy? Are you paying attention to me? Am I safe? Am I loved enough to matter? In one classic experiment, called the "visual cliff test," researchers put infants on a raised platform, a clear panel set in the middle. A baby crawling along the platform, looking down, would suddenly see a drop to the floor through the thick Plexiglas. The panel was as sturdy as the rest of the platform; but they didn't know that. Children would tremble there, fingers still gripping the opaque boards of the platform, staring down that steep virtual cliff.

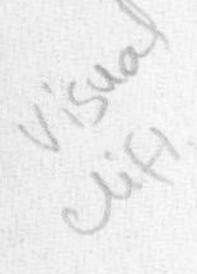

The children in this study were ten months old. They would reach the drop-off, hesitate, look down. Then they would turn their heads and look back at their mothers. The small sons and daughters would hold the edge while they studied their mothers' faces. If the mothers smiled and nodded, if their faces looked calm and encouraging, most of the babies went on over. A little tentatively maybe, their hands carefully feeling the slick surface of the Plexiglas. Sometimes the researchers told the mothers to wear a difference face. If the women looked fearful or doubtful, the infants' expressions began to mirror that. Their foreheads would wrinkle in apprehension. And then the babies would slowly back away from that perceived perilous edge. In psychology, the cliff experiment is justly famous. It stands as a stunning example of how much children look to their parents for answers—and receive them—without a word spoken.

The test is also a rare example of faith in another person. How many people in our lives trust us so much that if we nod and smile, they will chance a tumble down a cliff? So there's another point, here, about the specialized competence of infants. At this moment in their lives they give absolute trust. The same child, ten years later, relies on his own judgment, filtering a parent's assurances through experience. A brand new baby, who does not yet have such internal judgment, *must* rely on others. And since gathering facial information is imperative, babies *must* become adept at reading the subtle signals in a change of expression. They can use a mother's response to calm their own fears—or to validate them. They are like tiny treasure hunters, carefully searching the facial maps around them.

"Clearly, the emotional state of others is of fundamental importance to the infant's emotional state," says Harvard child psychiatrist Edward Tronick. His choice of the word *others* rather than *mothers* is deliberate. Children form many important relationships with adults. A "mother" may be biological, adoptive, guardian, foster, grandparent, relative, friend. In recognizing the full range of emotional connection and intimacy, our society has begun to embrace a closer role for fathers as well. In 1994, poet and science writer Diane Ackerman wrote that, compared to a mother's love, "a father's love . . . is more distanced, and often has conditions attached to it." Now, almost a decade later, our culture seeks to bring the father into that emotionally tight inner circle of the family. Infants may also scan a dad's face for comfort and for the kind of unconditional love that used to be seen as a mother's specialty. Of course, as Harry Harlow pointed out, the majority of infants in our world still have high hopes that mother will be there, smiling or frowning, when a potential cliff looms in view.

Babies send their parents nonverbal messages, too. Adults, though, aren't as adept at reading them. Some are easy enough. Infants smile when they are pleased; cling when they need contact; follow with their eyes when they are worried that we may leave. They cry when they want help or comfort—although exactly what they want can be tricky to figure out. If small children aren't reassured, if no one responds, they comfort themselves. In another study, psychologists placed a bright toy near a baby but just out of her reach. The infants in this study were tested individually—but they mirrored each other's behaviors anyway. The babies were too young to crawl to the ball. They tried, though. The scientists watched as the babies struggled to reach the bright ball, stretching out hands and failing, stretching and failing again.

In frustration, the infants sobbed to themselves. If help still failed to arrive, they would try to calm themselves down. They would suck their thumbs or look deliberately away from the toy. Thumb sucking turns out to be one of those natural resources, an effective way for babies to comfort themselves. Infants also calm themselves by the simple act of looking away. If a parent frustrates, if a toy rolls away and can't be reached, a simple way to cope is to focus on something else. It's a lesson learned in the first months of our lives that holds up well for the rest of our lives.

And that's exactly what babies do. An observant parent can see the child's eyes flick away—to a blanket, a wall, into the air even, but away from the source of unhappiness. If we—as parents—are paying attention, we may recognize this gaze-away as a message to us. The baby needs downtime. Even the smallest humans, the most dependent and connected, sometimes need resting space—the infant equivalent of Zen meditation or a walk alone in a quiet woodland.

What we parents won't see, of course, is the simple, lovely biology that runs stream-like through a baby's response to tranquility. Scientists have been able to track that internal shift in the most straightforward way. When a tired or frustrated baby looks away, her heart rate steadies and drops. If researchers have put a few sensors in place, they can see that change in the green line that indicates heartbeat. It's like watching water change at the sea front, from choppy little waves to smooth shiny swells. Thus the machinery of medicine can track the way the heart begins to ease.

Back in 1983, Ed Tronick at Harvard had begun to consider the power of this interaction between parent and child. It occurred to him that the I-smile-you-smile-back kind of relationship could be the basis of an interesting experiment. It wasn't the physical smile that interested him so much. It was what it represented—the give and give back between mother and child. When a toy is unreachable, an adult is instructed to respond in that experiment to the baby's signaling for help. The lab assistant will always eventually move the toy into those fat little hands. But what if nothing the baby did elicited a response? What if the toy was left to hover out of reach? What if he crawled to the edge of that cliff, turned, and got nothing from his mother—no gleam of encouragement, no sudden look of alarm? What if an infant could coo and call and coax and find that he has nothing in his box of social skills that will get him an answer?

It was in those questions that Tronick thought he saw a way to tug at the mother-child bond, the tie that Harry Harlow had considered so unbreakable. Tronick came up with what he called the Face-to-Face Still-Face Paradigm. He and a colleague, Jeffrey Cohn, asked the mothers of three-month-olds simply to go blank for a few minutes while looking at their children. The "still face" test demanded only that—a total lack of response. The mother had to present a face frozen into neutrality. No anger or threat. No humor or love. The all-important facial map would show nothing but emotionally empty terrain.

"The effect on the infant is dramatic," Tronick wrote in an early publication, echoing his own initial astonishment at the power of that still face. "Infants almost immediately detect the change and attempt to solicit the mother's attention." When a mother still refused to respond, babies tried self-comfort. They sucked their thumbs. They looked away. Then the babies tried again, just to see a little response. They'd reach for their best tools to engage their mothers. Infants would smile, gurgle, and reach. And, as ordered, the mothers would return nothing. The babies would comfort themselves again. Then they would try again. And again. Babies know this matters. They're stubborn about it. But after a while, confronted with only that blank face, each child stopped trying.

"I remember when I first did the still-face paradigm," says Tronick, who today heads the pediatric research division at Harvard Medical School. He is a tall, elegant man with silvery hair, brilliant blue eyes, and a habit of saying very precisely what he thinks. "I have a sequence of infant photographs from the first study. Pictures of a three-month-old reacting to a mother holding a still face. First, the baby is solicitous, trying to appeal to the mother, then he starts sucking his thumb, and then he just collapses, curls up in a corner.

"I said to people, look, it's like Spitz's babies; it's like the monkeys in Harlow's study. Look at this emotional reaction." With that perspective, Tronick suddenly found himself at the receiving end of yet another of those Spitz-Bowlby-Harlow reactions. The psychologists he showed the pictures to thought that what they saw couldn't represent emotion. It seemed to Tronick that his colleagues were almost personally uncomfortable with the idea that the connection between mother and child could be so strong. The notion that relationships could matter that much was unnerving. "And people just didn't want to see it that way. It's too close. I think part of the reason that rejection occurs is that there's a denial going on. People don't want to believe that a child could be so hurt—or that we could be so hurtful."

And that—the willingness to explore the worst of our nature as well as the best—is one of the things that Tronick came to admire in Harry Harlow. Here was a psychologist who never pretended, who was willing to look at even the uncomfortable result. If he thought it was right, he would fight for it. People used to argue to Harry that his lonely baby monkeys just needed more cognitive input, a richer environment, Tronick remembers. Psychologists would insist that the dysfunctional behavior of baby rhesus with cloth mom couldn't possibly have anything to do with emotional needs. "And Harry just refused to back down from his own interpretation, that it was social connection, that it was input from the mother that made the difference. Harry, even when he was doing extreme experiments, always saw the normal side, and that was connection. He was never confused about what mattered."

During his fifty years in psychology, Harry Harlow explored many research interests. His was never a one-track mind. He had an infinite capacity for curiosity, a compulsive need always to go himself one better. What didn't fascinate him? Harry was interested in the structure of the brain, the biochemistry of behavior, play, mental abilities, and sex differences. But mostly he brought all this together in an exploration of the whole tangled messy business of relationships. If you line up his major works—learning abilities, curiosity, baby care, mother love, touch, social networks, loneliness, stress, abuse, depression—they all fit together into pieces of a living puzzle. Harry believed, entirely, in the power and importance of relationships; and if one is to trace his impact on his field, one should not look at one study, one thought, but at the way the studies and thoughts fit together. In the end, Harry Harlow's vision of the nature of love was a sweeping one. His studies still stand, like bedrock, for psychologists who believe that love matters, that social connection counts, that we are defined as individuals, in part, by our place in the community.

"Relationships with a capital R," says Sally Mendoza, chair of the psychology department at the University of California-Davis. Mendoza did her graduate work under Gig Levine, during his time at Stanford. She is a calm, friendly woman with a brilliant smile, an infectious laugh, and a razor-sharp mind. Unlike Tronick, she is not a Harlow fan. Mendoza came of age in the rising feminism of the 1970s and finds it hard to like Harry's sarcastic and sometimes misogynistic style. But, even so, she has long believed that the way we connect is absolutely, fundamentally important in understanding ourselves—and any social species.

Even in graduate school, Mendoza was fascinated by relationships. Her idea was that we rarely act in isolation. Social connections influence many of our behaviors, underlie our decisions. Consider an observable behavior—from goofing with a friend to grieving over a lost lover. Mendoza was sure that each interaction was more than visible externally. It also changed internal physiology and chemistry.

Behind her idea lies a provocative theory: that our individual body chemistry is not so individual at all; that each of us is designed, in part, just to respond to the other people in our lives.

If so, then the lyric insistence of the seventeenth-century poet John Donne that "no man is an island" takes on a scientific literalness. We become inseparable from the fine fragile fabric of our relationships. "People told me I was crazy," Mendoza says. "I'd present this in an audience with people like Frank Beach [a pioneer in the study of hormones and behavior] and everyone would go after me, asking 'What's the mechanism? Are you saying that just relationships can have an independent effect?'

I'd say, 'Yes.'

They'd say, 'How?'

I'd say, 'I don't know' and they'd say, 'You're crazy.'

"But Harlow and Bowlby did have a big impact on thinking about relationships," Mendoza adds. Gradually, the field also recognized that her heretical notions might actually have some credibility. The power of those first cloth mother studies was inescapable, she believes. Who could deny the image of a baby monkey holding as if to a lifeline onto that artificially warmed terrycloth body? There's another study, out of the Harlow lab, that speaks even more to her. It's the Butler box in its "love machine" days. Mendoza could not set aside the image of the little monkey locked inside Butler's box, tirelessly opening a window for a glimpse of his mother. "And that's why I started reading Harlow. He completely strips away everything. Harlow's work tells you that without social support, you are in real trouble. You can end up in pathological personality development."

Our bodies know this; our brains recognize it subconsciously, even if we cannot accept it intellectually. Or so Mendoza suggests. We spend many of our limited waking minutes on each other. Even office life thrives on gossip and jokes and friendships. Parents with demanding jobs still huddle over homework with their children, cheer them at soccer games, fall asleep reading to them at night. Adult children still telephone their parents long after they no longer "need" them. We lunch, we date, we party, we spend quiet evenings at home; often the very best minutes of our days are the connected ones. And Mendoza believes that our particular biological nature demands this. If you think of the nature of love as a multifaceted gem of an idea, then our need to belong is a major facet. Without even thinking about it, "we spend a huge amount of time in relationships," she says. "That should tell us that it's inordinately important, that relationships are critical to biology."

No one tells Mendoza she's crazy these days. She works in the hot new psychology specialty called the biology of emotions. At the California Regional Primate Research Center, another of the NIH facilities created by Harry Harlow and his colleagues, Mendoza and Bill Mason, among others, have been trying to better define the brain anatomy and neurochemistry that helps sustain those bonds. Mendoza has looked at the intricate squirrel monkey society as an example. She finds that even peripheral relationships matter to these small, tightly networked animals. If Mendoza takes a squirrel monkey out of his group, she can measure a sudden spike in the animal's stress hormones. The rise isn't only in the separated individual. The hormone blazes across the group, even in monkeys who rarely spent time with the missing animal. Everyone registers that someone is missing. She suspects that we humans respond similarly to minor relationship changes—a coworker's leaving, a neighbor's moving on. It's a reminder that we weave our social fabric from many, many threads. There's a reassuring aspect to living in that complex of relationships. If one fails us, there are still others to keep the net stretched beneath us.

"One person may go to a single relationship for everything they need. I rely on a rich network of friends," Mendoza says. "And I firmly believe that you can make up for a nuclear family more easily than you can make up for friendships. Harlow saw that in monkey communities. The friendship, the peer relationships, the kin networks." There are different ways of describing Harry's idea that we need many "affectional systems" in our lives—friendships, as Mendoza says, community as Tronick says—the more the merrier, it takes a village, no man is an island. And Mendoza may be right; certainly Harry would have thought so, that one's nuclear family need not be the only—or even the best—family in our lives. We can and do extend our family circle with friendship and sometimes it's the extended part that matters the most.

"He was ahead of the curve," says Steve Suomi, "by at least thirty years. He was the first to look seriously at social behavior as it emerges in a developmental sense. He was interested in the layers of relationships, between mothers and infants, infants and other parts of the social world. His work preceded substantially the current argument over who is more important, parents or peers." As an example, Suomi cites the well-publicized 1998 book, *The Nurture Assumption,* which argued that peers and peer pressure could outweigh parents' influence. Like many researchers concerned about early childhood development, Suomi is wary of the book's message. He thinks the author, Judith Rich Harris, took the modern emphasis on the non-nuclear family to a risky extreme. It's true that his old mentor, Harry Harlow, believed that childhood friends mattered hugely, in part as a trial run for adult relationships. The playful days of early friendships do teach us some of the subtleties of building the social safety net. They can also buffer us against a dysfunctional family. Like Suomi, though, Harry would not have agreed that friendships make our first connections, mother-to-child, unimportant. His perspective was more complex than that.

In essence, Harry said, one good relationship opens the way to the next. There may be phases of our lives when friendships or partnerships seem more powerful than our original families. But our ability to make those later relationships may well depend on what each child gets from his or her parents. We learn about love and connection starting in the first microseconds of our lives. For better or for worse, those lessons last us a lifetime.

In their paper "Learning to Love," Harry and Peggy Harlow wrote that one outstanding quality of the good primate mother's behavior is "total or near total acceptance of her infant." In this model, the infant can do no wrong. The mother anxiously supervises his beginning sallies beyond the protective reach of her arm. She will scoop the child back if necessary. The baby, as a result, gains confidence in her protection and "total, tender, loving care." Harry agreed with that central tenet of attachment theory—if we don't have a secure attachment as a child, we may struggle throughout our lives to feel secure in all relationships. John Bowlby himself used to express great exasperation with the Western notion that dependency was a bad thing. Bowlby sometimes worried that we push our children away so fast, we value the model of independence so ridiculously much, that we rarely pause to acknowledge that dependence can also be both good and natural. A child depends on her caretaker, a dependency starts with simple survival and grows into real affection; and throughout our lives, we always depend on the affection of others. A part of any good relationship—child or adult—is the secure base, Bowlby argued, and if we are lucky, we may happily spend our lives exploring the world but never doubting our warm welcome at home. The best adult life, he once said, consists of explorations secured by a loving relationship—no different, really, from the young child on a playground who is fascinated by the new possibilities but still looking over her shoulder to see that mother or father is standing by. "On this foundation, it seems, the rest of [a child's] emotional life is built—without this foundation there is risk for future happiness and health," Bowlby wrote.

In rhesus monkey society, those first loving bonds are almost exclusively the responsibility of the female. And Mendoza was also ahead of her time in realizing that they resonate within. Scientists have learned ways to measure the internal biology of that relationship. There is singularly comforting body chemistry to being hugged by a parent who loves you. If a mother monkey scoops a baby close against her chest, heart rates drop—even more beautifully than when a baby looks away from frustration. Researchers have measured the same peaceful response in both boy and girl monkeys.

Their stress hormones drop, their entire systems seem to relax and smooth over. An identical reaction can be seen in human children. A child tucked against his mother's shoulder seems lulled into that easy chemistry of contentment.

It was this lovely image—Madonna and child in perfect tranquility—that helped foster the "Velcro-mother" idea, as critics called it, the notion that mother and child needed skin-to-skin bonding time after a birth. Two pediatricians, Marshall Klaus and John Kennell, proposed the idea in the early 1970s. It may seem simplistic today. At the time, though, there was a compassionate logic to it. Klaus and Kennell were justifiably exasperated by hospital rules. The two doctors crusaded against regulations that forbade mothers to keep newborn babies close by and parents to stay with sick children. In the matter of newborns, the pair suggested that real harm could come from such policies. Perhaps there might even be a critical minute, as it were, in which mothers had to be there, had to hold and cuddle. If a mother missed that moment, they argued, she might not bond with her child. Klaus and Kennell pointed out that although only 7 to 8 percent of babies born at the time were premature, from 25 to 41 percent of battered infants were carefully isolated preemies.

The pediatricians wondered whether the blame lay with hospital practice. Perhaps medical administrators were doing real harm. By enforcing separation, hospitals were causing mother and child to miss that all-important moment of connection. Klaus and Kennell wondered, for instance, whether mothers who were allowed to stroke and cuddle their babies loved them more. If so, they might be less likely to be abusive. It was a wonderfully appealing idea for many people. If they were right, of course, they could cure all kinds of dysfunctional families. The mother-child "moment" caught on almost instantly. Baby bonding videos flooded the market; in 1978, the American Medical Association made a formal statement in support of early cuddling. Of course, women who weren't able to snuggle, following general anesthesia c-sections, for instance, were tumbled into needless guilt and worry.

Needless because the mother-child relationship once again proved more complicated. As charming as it may seem to some, one near-magical bonding moment would make us even more vulnerable than we already are. A species such as ours, which must protect and nurture its young for years, would hardly be limited to developing love only in the first few minutes after birth. "Human infants are so helpless that it would be far more likely that bonding would be very flexible," says Meredith Small, a professor of anthropology at Cornell University and author of *Our Babies, Ourselves.* "It doesn't happen in half an hour. The connection to a child is a process.

"The good thing that Klaus and Kennell did is that they helped open people's eyes to the idea that the baby should be with the mother most of the time. So they were a little radical, sure. But sometimes you have to burn a bra to get people to look." It's not that connected minute that is so important. It's the stable and reliable connection. The emphasis on bonding at least reminded people that the best relationships begin early, and their beginnings are intense. The rest of the story is that the relationship needs to stay intense, perhaps for years to come. Children need attention in a long-term, not a short-term, sense.

When she lectures on childcare, Small still reinforces her position with Harry's mother-child experiments. She shows slides of the baby monkeys cuddling against cloth mom. She hasn't found a better visual example of the simple need to hold and be held. She wishes she didn't still need that evidence, that we already had learned the lesson. But, Small says, even today people argue against that weight of commitment. "American culture is built on individual achievement. You're told to be independent, self-reliant, get through life on your own. And that's in direct conflict with how humans are designed, evolutionarily and biologically. If you look at other primates, little kids and even adults are meant to be together. We're not like a bunch of wildebeests on the savanna. We're supposed to be dependent on each other, children especially. That's what all the evidence showed, that's what Harlow's work showed, that it's natural for the little rhesus to be connected to its mother. Being disconnected is like being punished."

With his usual direct approach, Harry also acknowledged the difficulties of balancing the needs of the child against the needs of the professional woman. "The working mother probably doesn't help the structure of the nuclear family," he said. "It is difficult for anyone to substitute for a mother." Who do you find who loves the child the way you do? If we are honest, the answer is usually no one. In Western culture, most of us no longer have someone from an extended family system to watch over our fledglings. We turn instead to paid day care and we promise ourselves that we will make it up to the child, in the evening and on the weekends. Sounding surprisingly twenty-first century, Harry Harlow said many years ago that such balancing acts are not unreasonable. He thought that if a mother carefully chose good childcare, and, when she came home, put in undivided time, children could still grow up with a strong sense of love and security. He cited Peggy's work on this point; the nuclear family studies, he told one newspaper reporter, made it clear that a mother is part—but not all—of a whole family support system. To raise a secure and emotionally healthy child, she needn't physically hover every minute—if there are other dedicated caretakers, if children grow up in a close network of friends and family. But when the mother comes home, Harry emphasized, she really needs to *be* there. No stacks of paperwork, no constant phone calls to colleagues. On evenings and weekends, he said, she should be a mother first, a company employee second or third, or somewhere even lower on the list.

In 1947, when Bowlby began making his case for motherhood, just 12 percent of mothers with young children (under the age of six) worked outside the home. In 1997, that number had risen to 64 percent. If we still believed in John Watson's dictate of distance between parent and child, there would be nothing troubling about those statistics. They might be said to indicate a healthy trend. But thanks to all the scientists who changed that perception—Spitz and Robertson and Bowlby and Ainsworth and Harlow and countless others—we worry that the distance is an emotional void. We worry that extended childcare is a social experiment. We worry about the risks inherent in any experiment. There looms the fear that we are raising a generation of children so loosely attached to their parents that they will become socially adrift. Of course, you could argue that if John Watson didn't engineer that disconnect, perhaps no one could.

In the early 1990s, the National Institute of Child Health and Development (NICHD) began a study to investigate the possibility of a link between children who spend much time in day care and children who are insecurely attached to their mothers. Does day care influence children to grow up without faith in the affection and security of home? The institute made a major commitment to this question. The study is still ongoing. It involves twenty-four scientists and 1,360 children from families fanned through a range of neighborhoods and income levels. The first research phase compared three groups of children: toddlers who stayed home; those who spent ten hours or less in child care centers; and those who were thirty hours or more a week in day care.

The early results, published when the children were three, were the kind that Harry always liked best. They served up answers that seemed to be built out of solid common sense. When the scientists looked at all the families, the parents, the children, the centers, added everything up, and watched the interactions, once again it was relationships that made a difference. What mattered was the connection between mother and child, the affection between a child and the other caretakers. If babies from a loving home with responsive parents went into day care, most seemed as secure going out as coming in. If a child had a strained relationship and indifferent home setting and then went into an equally indifferent day care setting, the emotional distance between mother and child often became wider—and colder.

There were some interesting—not entirely surprising—complications to this picture. The study also showed that it was usually the already securely attached mothers who worked hard to find good day care. The mothers with insecurely attached children tended to be less fussy about where their toddlers spent time. There was a kind of feedback system. The mothers of the more insecurely attached children didn't show the kind of protective behaviors associated with security. They didn't watch as Harry's good monkey mothers watched. They didn't pay attention the way a good Bowlby mother would. Their children—put into Ainsworth's strange situation test—didn't seek comfort the way most of the others sought it. When their mothers returned at day's end, as Ainsworth had consistently found, these children did not run to them in joy and relief. Neither did they coax affection from alternate adults. Ainsworth had also shown that mothers of insecurely attached children don't like to be touched. Growing up in those households, it seemed, their children had never learned the comfort factor of a cuddle.

You might ask whether relationships improved for the insecurely attached child who ended up in a terrific day care situation? Could regular days and warm, affectionate teachers tip that balance? Could the child learn to reach out more, repair his patchy support network? And the answer is another one that is less than perfect. Good day care didn't fix the relationship with mother. It could, however, improve other connections. Children did become easier with others, a little friendlier. That in turned helped them to build other relationships. Children could learn social skills from their caretakers that they might not learn at home. In that sense, affection, even in an institutional center, can help build a stronger foundation. As Bill Mason found with his mobile surrogates, we social species can draw a lot of benefit out of a very small amount of interaction and support.

Of course, big studies like this tend to project the big picture. Child by child, no two looked exactly alike. Not every dedicated parent had a securely attached child; not every harried and indifferent mother produced insecure attachment in her toddler.

In other words, attachment is complicated; in other words, Bowlby was often right and sometimes wrong. You can separate mother and child, even every day, and not break the bond. You can keep the child at home and still end up with insecure attachment—as the NICHD study also found. "Clearly, then, attachment theory is by no means without flaws, holes and huge unanswered questions," wrote psychologist and Bowlby biographer Robert Karen. "Various studies suggest that we cannot be as confident as we once were about the parenting styles that lead to [insecure] attachment or the degree to which inborn and cultural mores may also play a part." It might be that the child is irritable, Karen pointed out. Or that the mother is too nervous, overreacts to the baby's signals, ignores the look-away signal that says he needs downtime. Perhaps it's the baby who pulls back and not the mother. Perhaps, too, attachment theory is hopelessly mother-centric, demands too much of one parent, doesn't make room for the help of others. One question raised by the various mothers in Harry's laboratory was: Which is better: a bad mother or no mother?

One of the more interesting twists on "the right mother" or parent or guardian comes from Steve Suomi's research. Suomi was one of Harry's favorite graduate students. A stocky, fair-haired man, he brought to research the kind of single-minded intensity that Harry possessed himself and admired in others. Suomi considered a lifetime career at Wisconsin, after Harry had retired. He held a faculty position at the university for twelve years before the National Institutes of Health "made an offer I couldn't refuse" in the early 1980s. He still works at NIH, where he studies both monkeys and humans in his job as director of the Laboratory of Comparative Ethology in Poolesville, Maryland.

Suomi's study raised a deceptively simple question: Is the biological mother always the best mother for the baby? To evaluate parenting styles, he compared biological monkey mothers to foster monkey mothers. He chose with care. The foster mothers were "supermoms," picked for their nurturing style and—perhaps as a result of that style—their securely attached offspring. The comparison mothers were not rejecting or abusive. They were just a little less interested in their children, less devoted. The children in their care, though, were identical in nature. Both sets of mothers had to care for some high-maintenance babies, unusually nervous and jittery little monkeys.

The little monkeys simply did best with the most loving mother. Under tender loving care, the jumpy little monkeys grew visibly less stressed and, eventually, into nurturers themselves. "Those high reactive kids, reared by supermoms, now have kids of their own," Suomi says. "And they are supermoms themselves. It appears to be a nongenetic means of transferring behavior to next generation."

The little monkeys with their less engaged mothers did not show such a dramatic temperament change. The NIH researchers intensified the study. They selectively bred for highly charged monkeys. Those monkeys had super-charged children. Again, those infants were either kept with their high-intensity parent or placed with a loving foster parent. In this study, you could watch the nervous parent create the nervous child. The high-stress parents weren't unkind. They were just so jangled and distracted that it was difficult for them to really concentrate on the child. They were absorbed by jumping and responding and fretting. And so, it turned out, were their children. They were unnerved by the slightest change. They clung desperately to their mothers, apparently even afraid of inching away to explore. If the scientists provided new toys, altered the dinner menu, changed anything, the babies appeared instantly threatened. The cage would explode in a cacophony of alarm screeches—mother and child echoing each other in dismay. The difference in the foster families, thus, was almost deafening by contrast. There was plenty of conversation but not much screeching. There was no evidence that these babies had been nervous little infants when they were born. They grew up calmer, this time mirroring their foster mothers' personalities. The infants acquired other benefits from growing away from the natural nest. The cross-fostered monkeys were often adventurous little animals. They explored with energy, made friends easily. They were unruffled by small changes. Foster mother and child alike remained unfazed if served oranges rather than apples for dinner.

To frame the experiment in Bowlby's theory, the fostered infants also appeared to be unusually securely attached. When the monkeys were six months old, the scientists experimentally separated them from their foster mothers. The little animals were definitely stressed. They devised a coping strategy, though. They recruited friends. And they kept those friends—it appeared that they were just likeable monkeys. The babies raised by their own nervous and preoccupied mothers were—not surprisingly, if you think about it—insecurely attached. They were timid with others. Their shyness made them unusually slow to befriend others. They were more traumatized by separation. And they tended to live separately. The nervous monkeys raised by nervous monkeys tended to become loners. Social contacts were too much. They often dropped to the bottom of the monkey hierarchy.

Suomi tracked the young monkeys from both groups until they became parents themselves. The nervous little monkeys grew into nervous mothers, continuing the cycle. Despite being born with that same antsy biochemistry, the cross-fostered monkeys parented like their sweet-natured foster mothers. Clearly, the benefits of affection and kindness rippled right through to the next generation. Suomi's study, in part, provided another reminder that genes are not destiny. It reinforced that lesson from Harry's lab—that the mother we are born with is not always the mother we need. And again it supported Bowlby's belief that the best lives have a secure base at the center. "Whereas insecure early attachments tend to make monkeys more reactive and impulsive, unusually secure attachment seems to have essentially the opposite effect," Suomi wrote. He was talking about the monkeys in his study only, of course, but it's safe to say that both John Bowlby and Harry Harlow would have been comfortable in applying that lesson to the rest of us.

There are several reminders in that elegant NIH experiment: that we need not grow up to be our mothers; that we may not want to;

that it's not easy to change. And that it may be unfair to load all our expectations and needs onto one parent, anyway. With the best intentions in the world, one person may not be able—or intended—to give a child everything he or she needs. The extended family, even the right child care provider may be exactly what's needed.

The perils of depending too much on the one, the only relationship, are beautifully illuminated in yet another primate experiment by another one of Harry's former graduate students, Leonard Rosenblum of SUNY-Brooklyn. Rosenblum compared pigtail and bonnet macaques. Bonnet babies grow up in a kind of bubbling community of friendly females. Although the mother cares for them, they are also enveloped by the other adult females who help raise them. In the words of Rosenblum, the bonnet infants are both mothered and "aunted." Pigtails are mother-raised only. Their watchful female parents keep them very close to home. If Rosenblum took mother out of the home cage, both pigtail and bonnet babies wailed with fear and loss. But the bonnets then quickly went to their "aunts" as a coping strategy. A little pigtail had no one to seek for comfort. The baby would call for his mother. The small monkey would then lapse into depression, hunch over, refuse even to look at other monkeys. Watching baby pigtails, you could wish them a few of those aunts.

In her book *Mother Nature,* California anthropologist Sarah Blaffer Hrdy builds an image of the good mother very different from that 1950s lonely but devoted nurturer. The mother Hrdy has in mind is also fiercely protective of the child, of course, but sometimes she is just plain fierce. Hrdy would have us get rid of that milky Madonna stereotype. She reminds us that mothers are still women with passion, and ambitions, and, yes, interests beyond the child. And as long as we are getting rid of stereotypes, Hrdy points out, there's no reason to assume that human beings should function like pigtail macaques, each mother solely responsible for her young. Why shouldn't we be like bonnet macaques, connecting in that more giving community of aunts or uncles or cousins or grandparents? Why should we cast the social support net so very narrowly?

Harvard child psychologist Ed Tronick also wonders about the one-on-one bond, what Tronick calls a monotropic relationship. He and his colleagues studied Efe pygmy infants as a way of exploring other parenting arrangements. An Efe baby, for at least the first four months, spends more than half of her time with adults other than her mother. Friendly adults cycle through the baby's life. There may be five helpers an hour, depending on who has time to share. The resulting bonds appear to be almost communal. Babies clearly recognize their mothers and fathers, but they may also attach to several adults. Adults, in turn, may form close bonds with several babies other than their own. Hrdy calls this kind of shared care "allomothering." Her view of allomothering encompasses both the natural tribal version and the twentieth-century modern American version, which can be paid day care, done well, done properly, with affection and stability.

"It's an experiment that we've got running," says Meredith Small, author of *Our Babies, Ourselves.* "We have nonrelatives with the kids. It's okay if they become like an extended family. The really important issue is not whether the toddler is learning colors and how to read at age three, but does that teacher hug your kid?" If we aren't going to return to the closely linked extended family, re-create ourselves in the Efe model, perhaps we need to make sure that our day care centers are more like families than tidily ordered schools. Craig and Sharon Ramey, at the University of Alabama, have tested super-intensity preschool programs for children, mostly children from disadvantaged families who are likely to have highly distracted parents. Consistently, the children in those programs thrive. Ramey suggests that his prototype day cares—one to three ratio, lots of hugging and touching—are designed to mimic the extended family nature of human evolution. "Whether it's a child in the inner city kept inside for safety or whether it's an only child on a suburban two acres, the effect is the same," Ramey says. "We have to find ways of countering the isolation of the family."

We might also, in these more modern times, consider further emphasizing the role of the father. For all that he was, an unlikely champion of heart-to-heart fathering, Harry Harlow saw that possibility in his research. He was one of the first, wrote psychologist Joseph Notterman in *The Evolution of Psychology,* "to recognize the liberating function" of those shared abilities and the father's ability to "thereby share in the development of infant love." Harry wasn't a natural champion. He studied rhesus macaques, after all, a mother-centric species if ever one existed. But there's nothing that says that solo mothering is a bred-in-the-bone primate characteristic. It's not even a consistent macaque trait, if one considers those well-aunted bonnets.

Bill Mason and Sally Mendoza, at the University of California-Davis, have done some remarkable work with the South American titi monkey—as gorgeous a ball of fluff as ever perched on a tree branch—and found that titi females bond mainly to their mates. The females are not noticeably maternal. When titis have children, the males take responsibility for about 80 percent of the childcare. The father is the nurturing one, the caregiver. If the scientists lift the mother from the family temporarily, the baby shows a bare flicker of stress response. But if they take out the dad? The infant monkey's cortisol rises like mercury on a hot day. Still, even titis confirm Harry's famous point that we don't merely love the warm body that feeds us. The titi mother nurses her baby for the first few months, as in any lactating species. It's the dad who holds and carries the child. And it's the dad who is beloved.

Chuck Snowdon, now head of the psychology department at the University of Wisconsin, has been working with another South American species, the cotton top tamarin. Cotton tops are tiny, dark-eyed monkeys with white-tufted heads. They live in extended families that are not only closely related but also fully engaged in supporting each other. Tamarins form a social network that relies on each member to share in childcare duties. How do the babies fare under this team-handling approach? Brilliantly, it turns out.

Among the cotton tops, mother, father, aunts, and older brothers and sisters all pitch in to raise the infant. Mother is the milk provider, but the baby attaches to the member of the family who spends most time with him. "When you look at all the caretakers—mother, father, oldest brother—when the baby is scared, he runs to the one who does the most nurturing," Snowden says. Because they are wafted around in the group, the infants also receive a steady diet of attention. And if the mother turns out to be a not-very-good mother, bored and restless, the father or a brother will take over more of the baby duties. "So basically, what's happening is one member of a family is compensating for the behavior of another," Snowdon says. As in other species, there are mothers from whom, given a choice, a baby might want to be slightly separated. "There are restricting mothers, there are laissez-faire mothers," Snowdon explains. "But if we look in our family of tamarins, the multiple caretakers buffer the effect the mother has. So if you were unlucky enough to have a weird mother, you'd be buffered. Of course, if you had a brilliant mother, that would be buffered some, too."

As Harry's work showed all too clearly, and as some of us know all too well, there's no guarantee that you won't end up with a weird mother or a bored mother or even a monster mother. "If you're going to work with love," Harry said, "you're going to have to work with all of its aspects." One of the risks of the one-on-one attachment is that you could end up with a brass-spike mother and no one else to hold you. As Snowdon points out, there's a tradeoff. If you share in several caretakers, you may miss the advantage of getting the total attention of the world's best mother. But you are never as vulnerable to the spiked parent. "Maybe we're moving back toward more cooperative child rearing," he says, "and my belief is that this is better."

And maybe we are moving in that direction—or at least some of us are. One of the questions that arises, as one considers the variations in parenting across the primate world, is whether we humans are able to choose the direction. Are we such a flexible primate species that we can pick and choose among the best strategies of our monkey relatives? Or, like them, do we follow an inherent species pattern, intensively mothering like the pigtail macaque, delicately sharing out the responsibilities like the cotton top tamarin? How much room is there to negotiate one's way to becoming the best mother possible? Or to avoid becoming the worst one?

Once again, we are left with one of those imperfect and complex answers. Clearly, some cultures, such as the Efe, do indeed practice a cotton top tamarin approach to life. Clearly again, Bowlby's model was based more on the mother-first model of the rhesus macaques—not to mention those passionately imprinted greylag goslings. If you assume that the best clue to our basic biology is in the majority pattern, then you can't simply dismiss Bowlby as an artifact from a less egalitarian society. Culture to culture, we still look mostly like a mother-centric species. That doesn't mean that mother is the only option. But it should remind us that for human babies a central parent figure is absolutely, undeniably important. Someone in the family has to be paying full attention to that baby. Bowlby was right when he said those early attachments—to mother, father, or loving caretaker—are always among the most powerful influences in our lives. Where we may indeed be flexible is at the individual level, in paying specific attention to the needs of our own specific children. If that seems too small a beginning, there is plenty of research to assure us that even small gestures matter.

Consider one of the first and most deceptively simple results of Harry's cloth-mother tests: that babies crave a soft touch. Since that time, researchers have been trying to figure out why. Why would a terrycloth-towel-wrapped mother be night and day compared to a wire one? What in our biology makes contact comfort so critical to healthy development?

The scientist who did some of the first and best work on the basic chemistry of touch is Saul Schanberg of the department of pharmacology at Duke University. Schanberg started in a non–Harry Harlow way, by looking at rats. Schanberg found that when mother rats licked their babies, the action produced a cascade of much needed compounds, in fact, the growth hormones that produce normal body development. Remove the mother—remove the touch of her tongue, and the baby rats became stunted beings. Put the mother back into the nest and the babies gratefully began to stretch outward and upward. In another reminder of the basic mechanics of motherhood, Schanberg also found that you could—at least with rats—simulate the mother's lick with a wet paintbrush.

The Duke mother-touch studies fit smoothly into the evolutionary concept of Bowlby's attachment theory. Schanberg suggested that the intense response to touch alerts us to a primitive survival mechanism, one that probably exists in many species. "Because mammals depend on maternal care for survival in their early weeks or months, the prolonged absence of a mother's touch, more than forty-five minutes in the rat, for instance, triggers a slowing of the infant's metabolism," he wrote. If his mother was missing, the baby rat used less energy. That meant he consumed less fuel. And that meant he could survive a longer separation from the mother. All well and good as long as she wasn't gone too long. Once she returned, Schanberg says, "The mother's touch reverses the process, so that growth resumes at normal rates." The baby who huddles into his crib and the little monkey who curls up at the edge of her cage appear hopeless. But we should be aware that some of the huddling is just conservation. It is a curious mixture of despair and hope. As they hunker down, the young animals are waiting for their mothers to come home and for everything to be all right.

Myron Hofer, at New York University, also explored the power of touch by studying rats. Hofer was a genius at considering the mechanics of mothering. He would take the mother rat out of the nest and substitute her essential elements: warmth, milk, stroking with a brush, sound (recordings of her squeaks); he even pumped her odor into the cage. Hofer found that only touch made a difference in how the little rats grew. So he brought the mothers back into the cages.

There was one catch. He kept them under anesthesia, so there was no touch and nuzzle and lick. A mother's inert presence helped not at all. The babies continued to quietly shrink away.

Schanberg then went on to do a classic study with Tiffany Field, at the University of Miami. The two researchers went back to Klaus and Kennell's concern with preemies, but from a different angle. They weren't looking at whether the infants bonded through touch, just whether the babies physically needed the human contact. Field and a crew of graduate students went into one preemie nursery and simply touched the babies. They did this just for fifteen minutes, three times a day. The touching was very deliberate—slow firm strokes, the gentle stretching of tiny arms and legs. The stroked infants grew 50 percent faster than the standard isolated preemies. They were more awake and active. They moved more easily. A year later, on cognitive and motor-skill tests, they looked stronger and smarter than preemies left alone in the standard incubator. Touch therapy is now a routine part of hospital procedures for premature infants.

Field went on to head up the University of Miami's Touch Research Institute (TRI), where she conducted numerous massage therapy studies. The bottom line in all those studies was that touch is good for your health, your immune system, your sleep, your anxiety level, your life. Eventually, researchers discovered that touch could be an antidote to the painful effects of a still-faced mother; if she gazed blankly but also touched and stroked, the babies seemed to feel connected still. If the mother added touch, her infant would continue to respond, smile, and look back.

Recall Gig Levine's studies that found a small, interesting break from mother actually improved a baby rat's life? That effect has been found over and over and since his first surprising—and nearly rejected—research. Researchers have polished, refined, and better explained those inexplicable results. Robert Sapolsky at Stanford and Michael Meaney at McGill University in Montreal expanded on Levine's original three minutes of handling by increasing it to a fifteen-minute break. Two years later, they could still pick out the handled rats by their capable responses—their smooth easy reactions to a strange situation. Their comparison rats—oversheltered and unhandled—were easily startled and prone to rapid increases in corticosterone, a rodent stress hormone comparable to cortisol in humans and other primates.

Corticosterone—and, scientists suspect, cortisol as well—turns out to provoke some chemistry that can actually damage neurons, notably in the hippocampus, where memories are often processed. So handled rats—and, Sapolsky speculated, well-nurtured children—may grow into a healthier adulthood, complete with a brain that stays efficient longer. "Real rats in the real world don't get handled by graduate students," Sapolsky notes. "Is there a natural world equivalent?" He and Meaney decided to compare natural mothering styles. Surely, they reasoned, not all rat mothers raise their young with equal attention and care. They were exactly right, of course, and you would find the same thing in humans, monkeys, and just about any other species. "There's lots of natural variation in mothering," Meaney says, "from good, to not very good, to very bad."

By very bad, he doesn't mean physically abusive. He means unreliable, distracted, neglectful. Even baby rats need a mother who pays attention—licks and cuddles and feeds and protects. What Meaney suspected might be really important was simply what a mother does—or doesn't do—as part of the everyday routine. So he looked at mothers who focused on their young by devotedly licking and grooming them. He then compared those nurturing females to others who just couldn't quite stay interested in the little rat pups. Meaney found that the rat pups blessed with mothers who spent a lot of time caring for them had less of that simmering stress chemistry and, therefore, distinctly healthier brains.

In other words, what Sapolsky calls "this grim cascade of stress-related degeneration" can be slowed, or even stopped, by something as apparently mundane as a mother who pays attention. "It doesn't have to depart that far from normal to have profound influence on development," Meaney emphasizes. "You don't have to beat children, to compromise development." He did look at childhood stressors beyond the usual variation in mothering. To do that, Meaney turned to a tried and true Harry Harlow technique—isolating baby from mother. In one study, he collaborated with Emory University psychologist Paul Plotsky. They lengthened the separation from mother rat to three hours a day for the first two weeks of their baby rats' lives. "The most potent effect on stress reactivity that we can achieve is with maternal separation," Meaney says simply. The two psychologists found that these more severely separated rats grew into chronically stressed adult rodents. Plotsky described them as skittish. They were anxious in new situations. They tended to crouch in one place. "They stick to dark protected places like corners or tunnels," Plotsky says.

And as they hunched into a corner, the rats' stress chemistry soared. Outwardly, they were sitting as still as possible. Inwardly, everything was vibrating. Heart rate, blood pressure, blood glucose, adrenaline, noradrenaline, the whole stress system was ratcheting up. Even in their familiar cage, the separated rats stayed restless and unusually aggressive. In monkeys, Plotsky says, you can sometimes induce this kind of chemistry without physical separation. A little mental separation will do—the kind you get with a distracted and overbusy mother. If researchers put a mother monkey and her baby into an environment in which the mother had to forage constantly for food, worry about meals, she paid less and less attention to her infant. When these baby monkeys were tested later, as adults, they looked—in their stress responses, anyway—a lot like rats who had been separated from their mothers. They stayed always just a little frantic.

Does this transfer to the way we treat our own children? Yes and absolutely no. If we've learned anything since the Watsonian psychology of the 1930s, it's that rats are not, after all, a flawless model of human behavior. They don't build that intense face-to-face attachment in their mother-and-child relationships. But rat work certainly raises some reasonable questions about early environment and relationships. The monkey studies raise more questions. And there is evidence, as Plotsky points out, that early experience does sensitize circuits in human brains, especially if it is a stressful experience. "Infant organisms are learning machines," Plotsky says.

A research team at McLean Hospital, a psychiatric teaching affiliate of Harvard Medical School, led by psychiatrist Martin Teicher, has been using brain imaging technologies to compare people from a safe and protective family and those who grew up in an abusive one. In children from unhappy homes, the researchers have seen arrested development of the left hemisphere. That left side tends to be the hemisphere associated with happiness and positive emotions. The scientists have observed similar stunting in a structure called the *cerebellar vermis,* which is linked to emotional balance. Teicher and his colleagues suspect that the wild swings of stress hormones and neurotransmitters, responding to abuse, can subtly restructure the brain to create such differences. They also think the changes may mean that the individual is "wired" to superimpose hostility on an environment. "We know that any animal exposed to stress and neglect early in life develops a brain that is wired to experience fear, anxiety, and stress," Teicher says. "We think the same is true of people." He also cites the "seminal" work of Harry Harlow as a major influence on his modern, high-tech exploration of the influence of parent on child.

A person too prone to perceive a threat may be equally prone to overreact to the perception. "You can imagine how a child with a history of physical abuse, entering preschool, might get into considerable trouble and have difficulty making stable friendships if he or she tends to see a 'threat' where none exists," Plotsky says. The researchers who study the effects of early damaging environment on children, almost to a person, want to find ways to turn that around. As they better understand the biological damage done by abuse and neglect, they wonder whether that hard-won knowledge can be used to help those children. Can we undo what is harmful to us in childhood? Can we preserve what is best?

Recently, Meaney again focused on rats with indifferent mothers, females who weren't particularly interested in licking and grooming. By now, you'd predict that these rat pups were doomed to corrosive high-stress chemistry. You could also envision them as those neurotic adult rats, hustling aimlessly around their cages. Meaney tried two kinds of therapies. As Steve Suomi had, he gave the baby rats better mothers. Again, he found that an anxious baby rat given to a nurturing mother will change for the better, become less stressed and happier. In this newer study, though, Meaney was also interested in trying to help animals that don't have a chance for a better parent. So he put other stressed infants into an enriched environment. Several times a day, he took the baby rats out of their plain home cages and placed them in larger pens equipped with ropes for climbing, running wheels, wood blocks, and other rodent entertainment. This was Gig Levine's idea of "handling" taken to a newly sophisticated level. It worked, too. In response to that engaging playground, as the rats looked about with interest, their stress levels came down—and stayed lower. The rats—compared to those from similarly neglectful homes—were noticeably easygoing as they grew up.

There was a curious catch, though. The enriched playground wasn't nearly as effective at fixing the problem as having a better mother. When Meaney studied the brains of these newly calmed rats, he found their internal stress response was still set on a high anxiety level. The psychologists tracking those rats now suspect that they didn't actually correct the stress problems. What the enrichment program did was strengthen other parts of the brain, enough that the rats could compensate. In effect, the rest of the brain was able to stabilize the system. Paul Plotsky thinks of this as not so much a fix as a bandage. "When you improve the rats' behavior, are you correcting the initial problems or are you creating a patch?" Plotsky asks. "The answer seems to be, at least in some cases, you are creating a patch."

And perhaps, sometimes, the patch is the best we're going to achieve. So far we haven't figured out a way to rescue all children soon enough, to stop child abuse, to guarantee every baby a loving home. So far, there's no guarantee that we will. So perhaps we should put the energy into making some damn good patches for damaged children.

A few people have put the patch idea into direct practice. One is a neuroscientist, Bruce Perry, the outspoken chief of child psychiatry, at Houston's Baylor University. Perry argues that our biology is designed for a more complex social world than even a good nuclear family may provide. "Our current living systems are disrespectful of the brain's potential," he says. "It's unfair to expect one or two parents to provide all of the rich opportunity that our brain is seeking."

Perry has also tried enrichment approaches, touch therapy, dance, art, storytelling, and drama. By doing brain imaging, he's been able to see that such activities can help strengthen specific parts of the brain; for example, storytelling can build up the outer cortex, and play therapy can stimulate the limbic system, at the base of the brain. The children who benefit the most from this, he says, are neglected children. They haven't had anyone to play with them, stimulate them, and teach them how to interact with others. "You smile at your mother. She doesn't smile back. You want to be hugged. She's busy; she pushes you away. You ask a question and she doesn't look at you when she answers. And so you're taught that smiling gets you nothing, that people don't want to look at you, that you are unwanted," Perry says, and there is both sympathy and frustration in his voice.

Studies of neglected children find that often what they see is a still-face, no matter what the expression. When shown photographs of facial expressions, abused children often mistakenly see anger. Neglected children, too, often see nothing. Many of them lack the basic face reading skills, period. Happy, sad, furious? They just weren't sure what a face might be telling them. Of course, this makes complete sense. Who would teach them to read a face? The mother who had no interest in them? The father who wasn't there? No one had been there to teach them how to interact with another person.

Once again, it was back to cloth mom and her empty heart and empty head.

This was what Harry saw as the ultimate failure of cloth moms; socially, he said, "they have an effective IQ of zero." The stuffed surrogate could offer her children a warm body—but teach them nothing about a living one. The final tally of Harry Harlow's studies, and those that grew out of them, gives us "a body of knowledge about the devastating effects of social isolation and their extreme resistance to treatment. Many people still do not appreciate how bad the effects are," says psychologist Irwin Bernstein. Sometimes it seems that this is the hole in the dike, the chink in the armor, of our very successful species—our need not just to *be* loved, but to *feel* loved, when no one is guaranteed either.

A parent might not respond to a child for many reasons: depression, stress, weariness, drugs, alcohol, indifference. If a parent turns away consistently, Ed Tronick suspects that the child begins to see herself as ineffective and helpless. Perhaps it's worth repeating that we all—child and adult—need at least one relationship we can lean on without worrying about falling. And by definition, this means that both people in the relationship must do their part—asking for what they want, answering, talking, listening, reaching out, and reaching back. One leans when she is weary, one supports when he is strong. The still-face experiment is all about the baby's seeking the adult's response: Smile back at me, talk back to me, touch me when I reach for you. That means it's also all about the adult's paying attention. "The infants' message is that their mothers should change what they are doing," Tronick says. And the point—at least for the infant—is that if you are paying attention, you are indeed going to catch her when she unexpectedly falls.

What's important is not that the mother gets it right every time. No mother studied by any psychologist responded perfectly to her child in every instance. No psychologist who studies mothers thinks that perfection has anything to do with good mothering. It's fixing mistakes that matters—even just the willingness to try again. Tronick found that when infants are confronted by a mismatch—I asked you for this, you gave me *that*—the babies usually just signal again. And he has analyzed what happens next. Thirty-four percent of mothers in his study recognized the baby's need next time round. Another 36 percent nailed it on the third try. "Infants and their mothers are constantly moving into mismatch states and then successfully repairing them," says Tronick. He thinks of this as interactive error and interactive repair. The mother plays peek-a-boo until the baby is overexcited. He looks away, he stops smiling—it's a message. Stop. I need a break. The responsive mother breaks off the game, lets the baby cool down, corrects the mistake, and returns gently to the game or goes on to something else.

If one returns to the idea of the right parent, it may simply be a mother or father who doesn't give up on the child. No one gets an infallible parent. No one gets a perfectly secure base every minute of every day. We have built-in buffers for that, all those self-comforting actions that everyone needs occasionally. There is no requirement for angelic perfection in parenting. The requirement is just to stay in there. Harry's research tells us that love is work. So do all the studies that follow. The nature of love is about paying attention to the people who matter, about still giving when you are too tired to give. Be a mother who listens, a father who cuddles, a friend who calls back, a helping neighbor, a loving child.

That emphasis on love in our everyday lives may be the best of that quiet revolution in psychology, the one that changed the way we think about love and relationship almost without our noticing that had happened. We take for granted now that parents should hug their children, that relationships are worth the time, that taking care of each other is part of the good life. It is such a good foundation that it's almost astonishing to consider how recent it is. For that foundation under our feet we owe a debt to Harry Harlow and to all the scientists who believed and worked toward a psychology of the heart.

At the end, in Harry's handiwork, there's nothing sentimental about love, no sunlit clouds and glory notes—it's a substantial, earthbound connection, grounded in effort, kindness, and decency. Learning to love, Harry liked to say, is really about learning to live. Perhaps everyday affection seems a small facet of love. Perhaps, though, it is the modest, steady responses that see us through day after day, that stretch into a life of close and loving relationships. Or, as Harry Harlow wrote to a friend, "Perhaps one should always be modest when talking about love."

EPILOGUE

Extreme Love

For better or for worse, our self-perception is never animal free. . . . There is no escape: human behavior is always placed in this larger context of other behaving organisms.

Frans de Waal,
***The Ape and the Sushi Master,* 2001**

"IF YOU'RE GOING TO WORK with love, you are going to have to work with all its aspects," Harry Harlow once said. No one could have meant that more sincerely. Harry was unflinching in pursuit of love in all its incarnations. His research led from the best of mother love to the worst. He looked at families made joyfully close and families shredded apart. He measured kindness. He measured hopelessness. He charted life surrounded by affection and life stripped of all relationships. He explored emotional damage and he insisted on exploring emotional healing, as well. Harry described the arc of those studies as: love created, love destroyed, love regained. No American researcher before or since has sent young primates through such a range of love's terrain, from transcendent to treacherous.

If he had only explored love at its best, the golden nature of touch, say, a discussion of the moral and ethical issues raised by Harry Harlow's work might not be necessary. But in the same way that his results helped transform our understanding of love, his open-ended inquiries helped transform our sense of ethical and moral limits in such research. Can you imagine choosing to do his experiments on total isolation, to induce such grief in a baby monkey that he literally dies in your care? To design and build a monster mother that flings a clinging baby across a cage? Yes, those were extremes, even in Harry's lab. But should research go to extremes? One of the questions that now underlies Harry's work is this: What are we willing to pay for knowledge? How far into the ethically risky realms of research should we go in pursuit of a promising idea, a compelling question?

Harry Harlow never denied that animals suffered in his laboratory. He was equally forthright about why he could live with that. "Remember, for every mistreated monkey there exist a million mistreated children," Harry said. "If my work will point this out and save only one million human children, I really can't get overly concerned about ten monkeys."

But other people can get concerned. Other people can mind a lot about those ten lost monkeys. Many among the animal rights movement still mind; they remember Harry Harlow all too well. His name doesn't speak to them of love or friendship or the absolute imperative of relationships. They remember him as the man who tortured small and helpless animals. They want the rest of us to remember him that way, too.

Harry Harlow's legacy can seem paradoxical, bright and shadowed at once. His work helped change psychology for the better. We now take for granted the idea of holding our children when they are frightened, of treating them with affection. We accept that standing by matters. Being ready to comfort or listen or laugh—being willing to give as well as to receive—is fundamental in a relationship. We believe that, too. But our acceptance represents a revolution of sorts in the study of relationships. Both as a society, and as individuals, we now believe that our watch counts, that how we treat others shapes them and also shapes us.

Even the darkest of the Wisconsin studies—the motherless monkey work, the evil-mother studies—spoke to that recognition of one person's influence over another. Harry's studies are now woven into the treatment of child abuse. They played a role in illuminating the strength of the connection between a child and a dangerous parent. They made real the long-term effects of what was once considered a brief period in a child's life. The controlled studies that he conducted could never have been done in children. Early in the twentieth century, the National Institutes of Health did receive a proposal to isolate children for up to two years. It was, naturally, rejected. "Since that time, nonhuman experiments have told us what results this inhuman experiment would have produced," Harry wrote in an introduction to a 1971 psychology textbook. His experiments are still used to counter criticisms that human data—continuing studies of children in orphanages—are just circumstantial. Attachment researchers say they still sometimes rely on the cloth-mother work to answer those who say that a parent's touch doesn't matter. "I think Harry would be surprised to realize how important he is in clinical treatment," said his old friend William Verplanck.

And yet, Harry's work also casts an ethical cloud over the research itself. It *is* hard to dismiss the image of a baby monkey who desperately clings to his mechanical mother while she shakes him until his bones rattle. It's hard to think of the infant who calls and calls for a mother who will never come back. There are photos from Harry's lab of monkeys who have been released from long-term isolation. The images make you think that, melodramatic or not, the term "pit of despair" is apt. The animals look like survivors of a concentration camp: eyes blind with horror, arms still wrapped around themselves. Is it hard to look past those haunted faces? Some would say impossible.

Harry's darkest work can seem so very dark that even some of his fellow psychologists stand deliberately back from it. Some worry about being associated with such politically difficult studies. Some are troubled by the ethical implications. There are those who could wish away the sorrow and loss unmistakable in even the black-and-white images of Harry's work. If you had never heard of Harry Harlow before you opened this book, if you wonder how a psychologist who did so much pioneering work can seem so invisible only twenty years after his death, that transparency is due partly to the unease he yet stirs in his own profession. "There's no doubt that he's been considered politically incorrect," says psychologist Duane Rumbaugh, Harry's friendly competitor in the science of primate intelligence. "It was surprising to me how fast the citations dropped off after his death."

In her exploration of the psychology and science of parenting, *Mother Nature,* Sarah Blaffer Hrdy makes one reference to Harry's work. She describes it simply as "bizarre." The recent *A General Theory of Love* argues, as Harry did, that love in childhood shapes our brains—and therefore our futures. In outlining their theory, California psychiatrists Thomas Lewis, Fari Amini, and Richard Lannon acknowledge the power and importance of the Harlow experiments. They also acknowledge them as destined for "perpetual notoriety."

Robert Sapolsky, the primate researcher known for his explorations of behavior and social connection, expresses the paradox in his 1994 book on the biology of stress, *Why Zebras Don't Get Ulcers:* "These were brutal studies," he wrote of the Wisconsin experiments. The legacy of the research still resonates with tension. Animal activists ask why Harry's research was necessary. Or, as Sapolsky paraphrased their question, "Why torture baby monkeys to prove the obvious?"

The first answer is that the importance of love and connection wasn't obvious at the time. When Harry first began investigating the idea that babies need to be touched, he was going directly against the standard teaching of his time. The mainstream position was that babies get nothing from touch and everything from the hands that feed them. Yes, evidence from orphanages and foundling homes and hospitals suggested that these ideas were wrong. But scientists who based their arguments on those human examples, such as Bowlby and Spitz, were frequently dismissed as lacking valid data. When Harry Harlow began his mother love studies, it was as if psychology was poised to wake up into a world where intimate relationships mattered. To push the field forward, some hardheaded data was needed.

"It was a set of ideas just waiting for confirmation," says Bob Zimmermann, who worked with Harry Harlow on the first cloth-mother experiments. "The results of the project implied that mother love was critical for normal development. Of course, Freud said that, but in a different sense. Now the mother love did not have a sexual connotation." And, Zimmermann points out, outside the psychology community, the Wisconsin experiments dovetailed beautifully into real experience and basic common sense. "My daughter, who is a nurse, made a good remark when we were talking about the surrogate project. She said that probably every nurse who worked in a preemie unit was nodding their heads when they read about Harry's work, and saying 'See, I told you so, cuddling and rocking pays off.'"

Zimmermann believes that "the surrogate project opened up areas of research in human development that normally would not have been funded in the 1960s and 1970s. What government group would have approved a grant to test whether the cuddling, stroking, and rocking of premature infants would enhance their development?" In the insular world of psychology, it did take the animal research and the neatly designed experiments, the graphs and the charts, and the coolly ordered data, to turn the argument. The answers we call obvious today seem so, in real measure, because Harry Harlow conducted exactly those studies that some people now condemn.

Sapolsky acknowledges the power of the lessons from the Wisconsin laboratory under Harry Harlow. And yet, and still, he confesses to being dismayed by the later, grimmer studies: "To animal rights activists who would ban all animal experimentation, I unapologetically say that I am in favor of the use of animals in research and that much good has come of this particular type of research. To the scientist who would deny the brutality of some types of animal research, I unapologetically say things can go too far."

Harry's acidic public persona adds a particularly sharp edge to this discussion. If scientists perform such ethically troubling work, we like them to behave as if they were ethically troubled. We'll forgive them some of the hurt if they acknowledge it. We'll soften our accusations toward those who appear to see our point. Harry didn't do any of that. Perhaps he didn't see the need—or perhaps he didn't see the issue, either. He had, of course, that tin ear for political change. That was obvious in the ways he baited feminists. As the animal rights movement took shape toward the end of his career, he baited those activists, too—without a thought to the consequences. The following is from a newspaper interview with Professor Harry Harlow:

"I certainly don't like monkeys. Sure I've known a few who were very adept at tests. Sure, I kinda liked them, more or less. But, by and large, I just have no feeling for them—at all. I spent a summer taming monkeys, eight in large cages. I would go out and sit beside them. This is where I learned about monkeys. I got to the point where they were not afraid of me." There was one female monkey, Harry recalled, who escaped into the runway connecting the outdoor cages with those inside and refused to budge. "So finally I did one of the most incredibly stupid things. I smacked her right across the face. Now, if she hadn't been a friend of mine, she would have bitten me all to hell. Instead, she smacked me right back—the last time I was smacked by a woman."

The above quote, with all its attitude and humor and love of storytelling, is from an October 1973 story in the old *Milwaukee Journal* honoring Harry's retirement from Wisconsin. The writer, Robert Bonin, is obviously entertained. He notes his subject's love of tweaking the politically correct. He describes Professor Harlow as a "likeably charming pseudo-curmudgeon." It's difficult to like monkeys, Harry tells Bonin, because monkeys don't like humans. "In all honesty, I should perhaps say that I like monkeys because they've certainly done more for me than I have for them."

And Harry Harlow was equally frank in his scientific publications. Animal researchers have always tended to smooth over their experiments, use jargon to describe their work in more indirect ways. Instead of writing that the research animals were killed, they'll write that the experiment was "terminated." Sometimes they avoid using the word "animal" at all. They'll just call it a "subject." No one ever could have accused Harry of such gentle misdirection, or of using scientific terminology to buffer what he did. If an animal died, he said so. If an animal suffered, he said so. One of his later papers was titled, bluntly, "Induction of Psychological Death in Monkeys." He used the term "rape rack." He used the term "vertical chamber," but he made sure that everyone knew it was also a "pit of despair." He wrote of evil mothers and monster mothers and brass-spike mothers, even though the latter were equipped with bumps rather than spikes. He wanted people to notice what he did—in all the dimensions. Sackett recalled arguing with his professor over the terminology: "I begged him not to do that. I said, 'Maybe we should make this work sound a little less depressing.' And Harry replied, 'You know, I like to grab people's attention.'"

Former colleagues, among them Bill Mason, found themselves reading his papers with dismay. "He would write about his experiments as if he did them with glee," Mason says. "It made my flesh creep." Sapolsky wrote that the isolation studies were among the most troubling and haunting in the history of science and that Harry's descriptions made them seem even more so: "Harlow's scientific writing displayed a striking callousness to the suffering of these animals." As you might imagine, scientists weren't the only ones to recognize those qualities in the publications from the Harlow laboratory.

Five years after Harry's death, in 1986, biologist Martin Stephens, now a vice president of the Humane Society of the United States, published a ninety-five-page report devoted to the evils of maternal deprivation research. It's worth noting that instead of calling it "The Nature of Love," as Harry did his best-known talk on surrogate mothers, Stephens devoted much space to "The Nature and Extent of Suffering." Stephens gives Harry Harlow full credit for attracting his attention: "In a way, because of his eccentricity, Harlow invited criticism and attention. More than any other psychologist, he was responsible for psychology being singled out for attention and focus by animal protection groups."

This can seem wildly unfair to psychologists when, undeniably, their colleagues in other disciplines were far more brutal. Seymour Levine sometimes marvels that Harry's work drew so much attention from activists when many uglier experiments occurred during the same time. In 1957, when the Wisconsin lab was carefully comparing cloth and wire mothers, one notable rat experiment involved dropping unanaesthetized animals into boiling water to measure blood changes in response to shock and pain. Cats were used to study muscle atrophy. Their hind legs were pinned for more than three months until the tissue withered. In military research, dogs were blasted with radiation until, as the researchers noted, their skin crisped. Monkeys were shot in the head to measure rifle bullet impact, or in the stomach to study blunt abdominal trauma. "So was it Harry's work or did he just provide a good controversial target?" Levine wonders.

Until late in Harry's career, animal activists were remarkably respectful of research priorities. They accepted, as did many American citizens, that scientists simply knew best. They might complain, they might write outraged letters, they might lobby the government on behalf of animals. But they were polite about it. In the 1950s, the American Humane Society even supported laws requiring animal shelters to turn their animals over to research labs. (This apparently was a little too respectful; dissident members pulled out and formed the less compliant Humane Society of the United States and the Washington D.C.–based Animal Welfare Institute [AWI].)

Christine Stevens, founder of AWI, doesn't recall scientists' being at all respectful in return. Dismissive, contemptuous, hostile would be more accurate. Stevens particularly remembers receiving a letter from a national scientific organization calling her a "social pervert." She responded by politely lobbying harder for a new animal welfare act, one that would for the first time include lab animals. The law passed in 1966, with one particularly significant provision. Medical researchers had been so outspoken in opposing protection for lab animals that even members of Congress began to mistrust them. The

National Institutes of Health was not given responsibility for inspecting the laboratories it funded; the law instead created an inspection division at the U.S. Department of Agriculture (USDA).

In the next decade, though, it became obvious that lab animal care was not being overhauled. The USDA didn't really want to hassle researchers over a few unhappy cats and rats. Even after the law passed, Stevens was able to put together a list of lab cruelties that included starving dogs and injured cats left untreated. Many scientists of the time acknowledge that they felt no pressure to improve animal care in their laboratories. "There was just so little respect for the [animal] welfare movement in the U.S. that it offered little or no check on the moral resources of the researcher," says Harry's former graduate student, John Gluck, now a bioethicist.

In 1981, coincidentally the year that Harry died, activists ran out of patience. That year, a university student named Alex Pacheco went undercover in a monkey lab in Silver Spring, Maryland. Pacheco and a friend, Ingrid Newkirk, had just started an advocacy group so small that it almost seemed a club. It had twenty members. They named their group PETA, People for Ethical Treatment of Animals. A polite name for a group convinced that politeness accomplished nothing.

Pacheco had picked his target carefully. He chose a well-known primate researcher named Edward Taub. In Taub's lab, scientists were studying injuries to the nervous system. To do so, they surgically mimicked such damage. In a typical operation, Taub and colleagues would open a monkey's spinal cord and slice sensory nerve connections to numb the animal's arms and legs. Taub's ultimate goal was recovery. He was trying to find out whether an animal could lose all sensation in a limb and regain function anyway. If so, perhaps, medical procedures could be found to help people in similar distress, such as paralyzed accident victims.

In management style, the lab was nothing like Harry Harlow's system. At Wisconsin, Harry—following years of monkey hoarding—had been obsessive about maintaining the physical health of his animals. The cages were cleaned and everything in them was cleaned regularly, too, even the surrogate mothers. The animals were still given vitamins, shots, and fresh fruit. After learning so much about isolation, he and his graduate students had opted for contact comfort. They tried to house at least two monkeys in each cage.

Taub caged each monkey separately—partly to reduce potential injuries. As a result, the macaques showed all the classic isolation behaviors. They paced, rocked, clasped themselves, and—introducing injury anyway—chewed on themselves. And because their limbs were numb, the monkeys couldn't tell when they tore through their own skin. They became marked by bleeding sores. Further, there was no nonsense about cleanliness and fresh fruit. Even the USDA inspection reports agree that the Maryland lab was filthy. Cockroaches scrambled through the cages. (One scientific supporter of Taub argued that the insects provided the monkeys with protein.)

Pacheco photographed the animals in secret, gathered testimony from sympathetic lab workers, and reported Taub to the Montgomery County police. PETA also made sure that every newspaper and television station in the area had copies of those damning photos. As it turned out, the timing was perfect. The general public no longer trusted scientists quite so much. Readers and viewers of the news agreed that the lab was practicing animal abuse at taxpayer expense. In the outcry, the county brought animal cruelty charges against the scientist and the university.

During the court process, Taub lost his monkeys, his grant, and eventually the lab itself. Pacheco's strategy had worked. It was clear to animal activists that if they waited for the government to help, for scientists to care about their animals, they could easily wait until Hell chilled down. The sluggish official response and the active indifference of the scientific community had convinced the people at PETA—and other organizations—that to save animals they needed to fight dirty and fight now.

And they did. Labs were broken into. Animals were let lose. Files were destroyed, death threats made, fake bombs delivered, buildings splashed with blood, private homes picketed. Harry was long buried in Tucson, by then, but his students were all too alive and all too visible. Bill Mason was burned in effigy in front of the UC-Davis primate center. Jim Sackett's house was spattered with rotting vegetables, ashes, and the bodies of dead rats. That type of anger hasn't diminished over the passing years, either. In the spring of 2000, protestors marched to Sackett's home in the middle of the night and kicked his front door open, apparently just to prove to him that they could. Gig Levine was bombarded with hate mail; one letter threatened death for himself and his family, concluding, "You and your sadistic father figure Harlow are as sick and unethical and bloodthirsty as anyone convicted in the Nuremberg trials."

The fury over Harry's work started after his death. Sometimes, it seemed as if he had calculated that perfectly. "It's as if he sat down and said, 'I'm not going to be around in another ten years. What I'd like to do then is leave a great big mess behind,'" Bill Mason says. Sometimes it seems timed to a different agenda, that animal activists knew they could do a better job of picking on a dead man. It's too bad, says Steve Suomi, because Harry would have loved the fight: "He was a person who was used to being controversial and he would have taken them to the cleaners." Irwin Bernstein makes the same point. "Harry was targeted after his death. I've always thought of that as cowardice. He could have defended himself more than adequately in life."

Further, Bernstein says, animal activists purposely exaggerate Harry's sins—they also describe brass-spike mother as having barbed points when she had only blunt knobs. Critics make it sound as if Harry had put every monkey in his lab into isolation, when it was only a carefully small number. Animal rights organizations give no credit to how seriously he took the welfare of his own animals. Duane Rumbaugh recalls that Harry thought the NIH cage-size requirements for adult monkeys were too small and built cages larger than required by the federal government.

Steve Suomi points out: "At the time that Harry was doing his mother love studies, the *standard* for housing primates in captivity, be it in labs or zoos, was individual housing, in other words, partial social isolation, until Harry showed how devastating it really was. And it took a long time in some places—actually, most of NIH prior to my move there—before those standards were changed, usually over the strong protests of the veterinarians responsible for taking care of the captive monkeys and apes."

Harry's experiments—and his vivid descriptions of them—may have invited his critics to take on what he did. Still, there's no doubt that some of their complaints are built on revisionist history. We may wish that the researchers of the mid-twentieth century shared our social consciousness. But the ethical questions that we raise about Harry Harlow's research designs are ethical questions that occurred later. For much of his career—barring the last isolation and depression studies—Harry was squarely in the mainstream of how scientists regarded research animals.

It's worth considering the exceptions, perhaps because it's too easy to gloss over moral issues by simply consigning them to history. The extremes of the Harlow lab did trouble people, even at the time. Psychology professor Kim Wallen, at Emory University, was a graduate student at Wisconsin in Harry's final years. Although Wallen didn't study under Harlow, he recalls the rippling sense of unease that the later work produced. "The view among other researchers was that you didn't need to put a monkey in a pit of despair to socially damage him. And yet as long as NIH funded the research, there was very little you could do. And maybe more than that, I don't think the ethical issues were generally raised or seen as a general concern in the 1970s. That just wasn't the case."

Gary Griffin, now in administration at Waterloo University, was working on his master's degree in psychology in Harlow's lab at about the time the most severe isolation studies started. As he recalls, Harlow suggested that Griffin take on some of this work for his thesis. And he did. "We isolated the monkeys for three months, then looked to see how they'd developed socially, then we did six months of isolation." Why six months? "Just seemed like a natural check point."

The results were horrific; animals stumbling blindly around their cages, rocking themselves, chewing their skin open. Griffin began to hate what he was doing. "We achieved real devastation. They were difficult, painful studies for the monkeys and the people. Imagine any animal that you know a lot about, a cat or a dog, putting it in a cage for three months and allowing it no contact with anything. They survive but it isn't pleasant." Griffin was troubled by a system that condoned such experiments: "I mumbled to Harry about the system but he made it clear that he wasn't interested." Griffin continues to believe that Harry's work was important and that animal research is important and should continue. "But I personally don't want to do the work. There's value in the experiments, I don't regret being involved, but I've decided it's not for me."

John Gluck, more than any of Harry's students, has tried to explore the ethical dilemmas raised by the specters of Harry's final experiments. "Harlow's colleagues, me included, never challenged him on the ethics points," Gluck says, flatly and with regret. "The strengths of our spines were not sufficient to carry the weight of our professional goals and our conscience." Harry was not the kind of professor to encourage such discussion; Griffin wasn't the only student, either, who was discouraged. Gluck doesn't hold his old professor responsible for that environment: "I am just saying that access to the moral resources, like empathy, comes from a community that sustains this kind of reflection. Harry neither created that type of community, nor did one emerge in the laboratory."

"No one said stop," says Marc Bekoff. "But Harry Harlow was very famous and you don't tell famous people to stop."

Marc Bekoff is a professor of population biology at the University of Colorado in Boulder. He's also a scientist with a passionate belief that research must be moral and ethical in its treatment of research subjects, human and nonhuman. He works with the famed primate researcher Jane Goodall, trying to teach primate conservation. They co-founded Ethologists for Ethical Treatment of Animals (EETA), making the point that it isn't only outside activists who think that animal welfare counts. Goodall and Bekoff are writing a book together; they're also making other scientists really angry.

Or at least, that's the impression Bekoff gets from the researchers who walk up to him at meetings and scold him. "They should feel good that it's people like me and Jane criticizing them. We're not against research. We ask questions, we try to not let people hide behind the veil of science," Bekoff says.

He and Goodall were recently at an animal behavior meeting where she infuriated researchers by using the word "prison" for "cage."

Bekoff's voice has a shrug to it; well, he says, cages *are* prisons. And "we're all accountable for what we do." He's recently been writing angry editorials about a colleague who takes baby rats away from their mothers to test their stress response. The studies are, in fact, much like those conducted by Michael Meaney and Robert Sapolsky. "What bothers me almost more than Harry's first experiments is that they keep getting done all over again," Bekoff says.

Bekoff lectures on Harry Frederick Harlow in his classes, but in a way that would undoubtedly startle the subject. He doesn't teach mother love or the magic of a hug. Bekoff asks his students whether the community of science should have allowed Harry's surrogate work to be done. And if you conclude that Harlow should never have done that work—never have taken baby monkeys from their mothers, caged them with air-blast mom, dropped them into vertical chambers—then, Bekoff says, the question of why such work continues becomes even more of an ethical dilemma.

"I find that Harry Harlow himself is not the major problem," Bekoff adds. The work is over, it's done, you can't get the monkeys back. "But social deprivation falls into a category of work that should never be done again, with all respect to Harry Harlow, and even though he did not make a mistake in his own eyes, we do not need to keep repeating this."

Martin Stephens found that isolation work peaked between 1965 and 1972. In those seven years, more than one hundred studies isolating lab animals—not just monkeys but dogs and cats—were reported. The Harlow lab conducted nine of the studies cited by Stephens, which made it hard to argue that Harry was solely responsible for the whole world of mother-child separation experiments. Harry's real sin, in Bekoff's eyes, is that he gave the experiments a kind of power and legitimacy that keeps them going today. "I could spend my life damning Harry Harlow, but where would that get me?" Bekoff asks. "I'm looking for institutional change, proactive change, and right now what I see is that he's a consciousness-raising tool."

The problem, Duane Rumbaugh says, is that animal activists have tunnel vision about the ethical issues raised by Harry's work. Yes, it's important to ask whether the work should have been done. But there's another set of ethical dilemmas that rise out of Harry's work, dilemmas that are equally important, equally troubling. And in Rumbaugh's opinion, these other issues aren't getting the attention they deserve. Monkeys are smart animals, really smart. Back in the 1930s and 1940s, Harry's work with the Wisconsin General Test Apparatus (WGTA) demonstrated that as emphatically as his cloth-mother studies would later make the connection between touch and love. And studies of primate intelligence have also gone far beyond the WGTA results. Rumbaugh himself has been instrumental in doing those studies across a range of species. He pioneered studies showing that chimpanzees could master the symbolic aspects of human language. He's shown that rhesus macaques can do simple math problems, play computer games—and even outscore their human trainers in shooting down digital targets on a screen. "The classic WGTA underestimates the rhesus by a 1,000 percent," Rumbaugh declares. "I'm really sorry that Harry wasn't alive when we made those discoveries. He would have been ecstatic."

But, Rumbaugh adds, our own society is still coming to terms with the bigger ethical questions raised by such discoveries. Should we conduct research on animals who are so smart, so socially complex, so closely related? In the primate family tree, rhesus macaques sit an uncomfortably narrow genetic distance from humans. Scientists estimate they share about 92 percent of our genes, and you can argue that their sometimes astonishingly human-like capabilities—from curiosity to game playing, from mothering to friendship—may reflect that linkage. Shouldn't we then question the morality of caging and experimenting on our kin? It's easy to judge Harry in hindsight, Rumbaugh says, and it will be easy for others to judge us in the same way. "Harry was a captive of his times, as are we," he says. "We, too, will be looked upon by future generations of scientists as less than sophisticated, less than human, less than sanguine. And they will be right. Of course, the generations that follow will hold the same of them."

By this reckoning, you could also argue that Harry Harlow's work helped build the platform on which animal rightists now take their stance. He greatly added to our appreciation of the intelligence and the social complexity of other primates. His studies, directly and indirectly, helped create that sensitive social consciousness that we value today. Rumbaugh doesn't bother to deny Bekoff's complaint that science can seem to repeat itself endlessly. It does repeat, sometimes for no good reason and sometimes for the best of reasons. Repeating an experiment, confirming a finding and improving on it, is a fundamental part of the scientific process. In considering the moral implications, though, we might weigh other reasons for repetition. Perhaps scientific research is sometimes redundant because we are slow to get the point. Perhaps humans need redundancy because we have to hear something over and over before we learn it—or accept it. Long ago, Harry himself made the comment about our understanding of love, that even God had to accept that we learn at our own rate.

Should we be angry with Harry or with ourselves for being such very slow learners? Perhaps it takes the extreme example of the isolated monkey or the baby in the box to force us to see the right and the wrong. Robert Sapolsky raises that point eloquently in *Why Zebras Don't Get Ulcers* when he considers the human species as it plods toward an understanding of affection: "It is sad and pathetic when we must experiment on infant animals in order to be taught the importance of love. But it is sadder and more pathetic to consider that we have learned about love so poorly and still have to be reminded of its importance at every opportunity." Living, loving, and learning are the most important parts of life, Harry Harlow wrote shortly before he left the University of Wisconsin. Learning never comes easy. And love is more difficult still. Yet we keep trying, those of us who have an inkling of what we're seeking. One more time, we tell ourselves, and perhaps we'll find the way.

The path to wisdom isn't well marked. There are plenty of signposts, but they're confusing, contradictory, humbling. So we turn to guides, those who can see a bit more clearly through the thicket, a bit farther into the distance. Harry Harlow—dispassionate, curious, and fearless in inquiry—was one of those guides. As objectively as he knew how to be, he underscored what should have been obvious, he insisted that good research should make sense, even on the emotional level. He wasn't perfect in the way he went about his work. It's impossible to like everything he did or the way he did it. In his zeal to explore even the ugly aspects of love, his experiments became ugly. Harry performed experiments that no one today should repeat. If ever there was a legitimate scientific need to put baby monkeys into vertical chambers, that need is past. Let us agree with Bekoff on this one. Once is more than enough.

But since we are so ridiculously slow, sometimes, to understand the lessons of love, perhaps we need to listen even to the most painful messages. No one who knows Harry's work could ever argue that babies do fine without companionship, that a caring mother doesn't matter, that we can thrive without ever being scooped up into someone's arms and reassured that the day is going to be all right. And since we—psychology as a profession, science as a whole, mothers and fathers and all of us—didn't fully believe that before Harry Harlow came along, then perhaps we needed—just once—to be smacked really hard with that truth so that we could never again doubt. Let us remember the best of Harry's contributions as well as the worst. Let us not slip backwards, ever, into believing that we are not necessary to each other's health and happiness. You don't have to like the way Harry found his answers. Almost no one could admire every choice he made. But neither should we pretend that he did anything less than arrive at some fundamental truth. Our challenge is not to squander it.

Notes

This book is based on a variety of sources—interviews, correspondence, Harry Harlow's unpublished memoirs/journal, books, magazines, newspapers, and research journals. Unless otherwise indicated in these notes, the comments of Harry Harlow's colleagues, students, and family members are based on direct interviews. Many people interviewed are not quoted in the text and yet their comments and perspectives did help shape the story, and contribute to the portrait of Harry Harlow, his family, and the University of Wisconsin-Madison's psychology department and primate laboratories during the time that he worked there. In that regard I would like to recognize the help given by: Leonard Berkowitz, professor emeritus of psychology, UW-Madison; and Harry's former graduate students and colleagues and friends: Dan Joslyn, Robert Bowman, Robert Dodsworth, John Bromer, Albert Erlebacher, Billey Levenson Fink, Leslie Hicks, Kenneth Michaels, Gil French, Carl Thompson, Arthur Riopelle, Joyce Rosevear, Bill Seay, Frank Farley, Brendan Maher, Greg Oden, William Prokasy, Eleanor Schmidt, and Marge Harris. The information they provided was invaluable to getting this story right.

Prologue: Love, Airborne

"The Measure of Love," *Conquest,* produced by the Public Affairs Department of CBS News; producer, Michael Sklar; writer, S. S. Schweitzer; director, Harold Mayer; aired November 1, 1959. Ernest R. Hilgard, *Psychology in America: A Historical Survey* (San Diego: Harcourt Brace Jovanovich, 1987).

Harry Harlow on proximity, quoted in John P. Gluck, "Harry Harlow: Lessons on Explanations, Ideas and Mentorship," *American Journal of Primatology* 7 (1984): 139–146.

Chapter One: The Invention of Harry Harlow

Susan Fulton Welty, *A Fair Field* (Detroit: Harlow Press, 1968); Charles J. Fulton, *History of Jefferson County, Iowa* (Chicago: S. J. Clarke Publishing Company, 1914); Robert P. Long, *Homegrown: An Iowa Memoir* (self-published, 1988); Susan Fulton Welty, *Man of Medicine and Merriment* (Rockton, Ill.: Basley Prington, 1991); *The Quill,* Fairfield High School Yearbook, 1913–1926. Harry's nephew, who also goes by the family name of Robert Israel, has archived a family history and photos at the Fairfield Public Library. It includes the photos of Harry's parents and an Israel family tree that goes all the way back to seventeenth-century England.

Harry Harlow's recollections of Fairfield are from unpublished journals/autobiography, courtesy of Robert Israel. Family history and photos are archived at Fairfield Public Library. Lon Israel's work records are from the Fairfield City Directory and from Harlow's unpublished descriptions. There are also some excellent descriptions of Harlow's childhood in W. Richard Dukelow, *The Alpha Males: An Early History of the Regional Primate Research Centers* (Lanham, Mass.: University Press of America, 1995).

Harry Harlow describes some of his early educational experiences in the book he co-edited with Clara Mears Harlow, and in "Birth of the Surrogate Mother," a chapter in *Discovery Processes in Modern Biology: People and Processes in Biological Discovery,* ed. W. R. Klemm (Huntington, N.Y.: R. E. Krieger Publishing Co., 1977). There are similar stories—as well as childhood background—in "The Evolution of Harlow Research," an introduction written by Clara Mears Harlow, *Learning to Love: The Selected Papers of H. F. Harlow,* ed. Clara Mears Harlow (New York: Praeger, 1986).

Background on Stanford and its psychology department: Annual Reports of the President of Stanford University, 1923–1931; Margaret Kimball, *Stanford: A Celebration in Pictures* (Stanford: Stanford University Press); *Jane Stanford's Inscriptions* (a publication of Stanford Memorial Church); correspondence from the Stanford archives between Lewis Terman and university officials. Background on Walter Miles and Calvin Stone in the *Reports,* an annual Stanford publication, which included activity summaries written by department heads. Lewis Terman filed the reports from 1924–1930, the period that I surveyed; in Ernest R. Hilgard, *Psychology in America: A Historical Survey* (San Diego: Harcourt Brace Jovanovich, 1987); and in John A. Popplestone and Marion White McPherson, *An Illustrated History of American Psychology* (Akron, Ohio: University of Akron Press, 1994); and from Harlow's personal recollections. The story of Calvin Stone and the rat bite incident is from Carol Tavris, "Harry, You Are Going to Go Down in History As the Father of the Cloth Mother," *Psychology Today* (April 1973).

Background on Lewis Terman from Joel Shurkin, *Terman's Kids: The Groundbreaking Study of How the Gifted Grow Up* (Boston: Little, Brown and Co., 1992); Henry L. Minton, *Lewis M. Terman: Pioneer in Psychological Testing* (New York: New York University Press, 1988); and from the Archives of the History of American Psychology, archived correspondence between May V. Seagoe and Harry Harlow, Nancy Bayley, Jessie Minton, and Robert Bernreuter concerning Seagoe's biography of Terman, *Terman and the Gifted* (Los Altos, Calif.: William Kaufmann, 1975). Terman's research is also thoroughly discussed in Hilgard and other history of psychology texts, including John A. Popplestone and Marion White McPherson, *An Illustrated History of American Psychology* (Akron, Ohio: University of Akron Press, 1994); C. James Goodwin, *A History of Modern Psychology* (New York: J. Wiley, 1999). For further background on the mixture of politics and science that encircles IQ testing, I used Mark Snyderman and Stanley Rothman, *The IQ Controversy* (New Brunswick, N.J.: Transaction Publishers, 1990). The story of Harry Harlow's name change is cited in Richard Dukelow, *The Alpha Males: An Early History of the Regional Primate Research Centers* (Lanham, Mass.: University Press of America, 1995); in Tavris's *Psychology Today* story, in Clara Harlow's brief biography of her husband, and in Harry's unpublished journal. Walter Miles's letter to Harry's father is archived at the library of the Wisconsin Regional Primate Research Center. Other context is provided by interviews with William Mason and Dorothy Eichorn.

Chapter Two: Untouched by Human Hands

Records on the early history of foundling homes from Sarah Blaffer Hrdy, *Mother Nature: A History of Mothers, Infants, and Natural Selection* (New York: Pantheon Books, 1999), and from William P. Letchworth, *Homes of Homeless Children: A Report on Orphan Asylums and Other Institutions for the Care of Children* (reprint, New York: Arno Press, Inc., 1974). Henry Chapin's paper is discussed in Robert Sapolsky, "How the Other Half Heals," *Discover*, vol. 19, no. 4 (April 1998). Chapin himself coauthored seven editions of a book on the subject. The first was Henry Dwight Chapin and Godfrey Roger Pisek, *Diseases of Infants and Children* (New York: Wood, 1909). (Chapin's report on infant deaths in institutions was published as "A Plea for Accurate Statistics in Infant's Institutions," *Journal of American Pediatrics Society* 27 [1915]: 180.)

L. Emmett Holt, R. L. Duffus, and L. Emmett Holt, Jr., *Pioneer of a Children's Century* (Appleton, London), 295; in Robert Karen, *Becoming*

Attached: Unfolding the Mystery of the Infant-Mother Bond and Its Impact on Later Life (New York: Warner Books, 1994); in Robert Sapolsky, "How the Other Half Heals," *Discover,* vol. 19, no. 4 (April 1998); in Robert Sapolsky, *Why Zebras Don't Get Ulcers* (New York: W. H. Freeman and Co., 1994); and in Sylvia Brody, *Patterns of Mothering* (New York: International Universities Press, Inc., 1956).

H. Arthur Allbutt, *The Wife's Handbook: How a Woman Should Order Herself During Pregnancy, in the Lying-In Room, and After Delivery: With Hints on the Management of the Baby, and on Other Matters of Importance, Necessary to Be Known By Married Women* (London: R. Forder, 1888).

Descriptions of childhood medical wards in Harry Bakwin, "Loneliness in Infants," *American Journal of Diseases of Children,* vol. 63 (1942); in Harry Bakwin, "Psychological Aspects of Pediatrics: Emotional Deprivation in Infants," *Journal of Pediatrics* (1948); in Robert Karen, *Becoming Attached: Unfolding the Mystery of the Infant-Mother Bond and Its Impact on Later Life* (New York: Warner Books, 1994).

Cooney's work is described in Marshall H. Klaus and John H. Kennell, *Parent-Infant Bonding* (St. Louis: C. V. Mosby Co., 1982). Another excellent summary can be found in Jules Older, *Touching Is Healing* (New York: Stein & Day, 1982).

Concerns that mothers don't want to touch their children are outlined in C. Anderson Aldrich and Mary Aldrich, *Babies Are Human Beings Too* (New York: The Macmillan Co., 1938).

Brenneman cited in Bakwin, "Loneliness in Infants." Background on Watson: Kerry W. Buckley, *Mechanical Man: John Broadus Watson and the Beginnings of Behaviorism* (New York: Guilford Press, 1989); James T. Todd and Edward K. Morris, eds., *Modern Perspectives on John B. Watson and Classical Behaviorism* (Westport, Conn.: Greenwood Press, 1994); and David Cohen, *John B. Watson, the Founder of Behaviourism: A Biography* (London and Boston: Routledge & Kegan Paul, 1979).

Watson and Stanley Hall's perspectives on parenting also discussed in Ernest R. Hilgard, *Psychology in America: A Historical Survey* (San Diego: Harcourt Brace Jovanovich, 1987); and C. James Goodwin, *A History of Modern Psychology* (New York: J. Wiley, 1999). The receptive audience of parents is discussed in Ann Hulbert, "The Century of the Child," *The Wilson Quarterly* (Winter 1999); and in Kim Klausner, "Worried Women: the Popularization of Scientific Motherhood in the 1920s," published on the History Students Association Home Page of San Francisco State University (http://www.sfsu.edu/-has/ex-post-facto/mothers.html) and explored in

Molly Ladd-Taylor, ed., *Raising a Baby the Government Way: Mothers' Letters to the Children's Bureau, 1915–1932* (New Brunswick, N.J.: Rutgers University Press, 1986).

Hospital policies discussed in the books of Klaus and Kennell and of Jules Older. The Minnesota "Child Care and Training" books were published by the Institute of Child Welfare, University of Minnesota Press, Minneapolis; I surveyed editions starting in 1929 and continuing through 1943. The publications of the federal Child's Bureau from 1914–1963 are reprinted in *Child Rearing Literature of Twentieth Century America* (New York: Arno Press, 1973). William Goldfarb, "The Effects of Early Institutional Care on Adolescent Personality," *Journal of Experimental Education,* vol. 12, no. 2 (December 1943); William Goldfarb, "Variations in Adolescent Adjustment of Institutionally-Reared Children, *American Journal of Orthopsychiatry* 17 (1947). Levy and Bender profiled in Karen, *Becoming Attached.* Issues of child isolation discussed in David M. Levy, *Maternal Overprotection* (New York, Columbia University Press, 1943).

Spitz and Katherine Wolf appear in Sheldon Gardner and Gwendolyn Stevens, *Red Vienna and the Golden Age of Psychology 1918–1938* (Praeger: New York, 1979). The work of both Spitz and Robertson is beautifully described in Karen's *Becoming Attached.* Karen's book is also, of course, a biography of John Bowlby, and includes a detailed discussion of his battles with Freudian psychiatry.

Of Bowlby's writings, I relied primarily on his three-volume series: John Bowlby, *Attachment and Loss,* 2nd ed. (New York: Basic Books, 1982), and *A Secure Base: Parent-Child Attachment and Healthy Human Development* (Basic Books, 1988). Specific articles included: John Bowlby, "Maternal Care and Health," World Health Organization (WHO) Monograph 2 (Geneva: 1951); John Bowlby, "The Nature of the Child's Tie to His Mother," *International Journal of Psycho-Analysis* 39 (1958): 350–373; John Bowlby, "Grief and Mourning in Infancy," *The Psychoanalytic Study of the Child,* vol. 15 (1960).

Both Karen's and Hrdy's books provide an excellent look at Bowlby's work and its influence. For a scientific overview of the field, I used Jude Cassidy and Phillip R. Shaver, eds., *Handbook of Attachment* (New York: The Guildford Press, 1999).

Although the tensions between Bowlby and Freudian psychiatry are discussed in the books cited above, I also consulted Edward Shorter, *A History of Psychiatry* (New York: John Wiley and Sons, 1997); and Harry K. Wells, *Sigmund Freud, A Pavlovian Critique* (London: Lawrence & Wishard, 1960).

Chapter Three: The Alpha Male

Harlow's first experiences at Wisconsin are detailed in his published memoir, "Birth of the Surrogate Mother," *Discovery Processes in Modern Biology*, ed. W. R. Klemm (Huntington, N.Y.: R. E. Krieger, 1977); and in Clara Harlow's introduction in *Learning to Love: The Selected Papers of H. F. Harlow* (New York: Praeger, 1986).

The descriptions of Clara Mears and many of her comments—here and throughout the book—are drawn from the questionnaires she filled out for Lewis Terman, now archived at the Stanford University psychology department, and from correspondence in those files between Clara, her mother, and Terman.

Gordon Allport's rebellion against rat research is discussed in *Rebels Within the Ranks: A Psychologist's Critique of Scientific Authority and Democratic Realities in New Deal America*, Katherine Pandora, (Cambridge: Cambridge University Press, 1997). Allport's career is also outlined in Ernest R. Hilgard, *Psychology in America: A Historical Survey* (San Diego: Harcourt Brace Jovanovich, 1987).

See previous chapter for sources on John B. Watson. Also, Roger R. Hock writes about Little Albert in *Forty Studies That Changed Psychology: Explorations Into the History of Psychological Research*, 3d ed. (Upper Saddle River, N.J.: Prentice Hall, 1999); G. Stanley Hall wrote about the goals and failures of psychology research in an editorial in the *American Journal of Psychology* 7 (1985): 3–8. These writings and other early landmarks in psychology can be found in "Classics in the History of Psychology," an Internet resource (http://psychclassics.yorku.ca) developed by Christopher D. Green, York University, Toronto.

Sechenev and Pavlov are discussed in C. James Goodwin, *A History of Modern Psychology* (New York: J. Wiley, 1999); and so is B. F. Skinner, who is also discussed in depth in Hilgard, *Psychology in America.* Skinner also wrote a two-part autobiography; I used the second part in researching this book: *The Shaping of a Behaviorist* (New York: Alfred A. Knopf, 1979). Both he and Pavlov are profiled in Hock's *Forty Experiments.*

Harry Harlow's descriptions of his efforts to launch an animal research program, including the cat and frog research, are from his unpublished memoirs. The story of Maggie and Jiggs can be found in "Birth of the Surrogate Mother" in the W. M. Klemm book, as can a discussion of Tommy the baboon.

Abraham Maslow's time at Wisconsin is discussed in Edward Hoffman, *The Right to Be Human: A Biography of Abraham Maslow* (Los Angeles: J. P. Tarcher, 1988); and in Richard J. Lowry, *A. H. Maslow: An Intellectual*

Portrait (Monterey, Calif.: Brooks/Cole Publishing Co., 1973). Harry Harlow discusses his admiration for Maslow in the 1973 *Psychology Today* interview with Carol Tavris and in correspondence with Maslow's wife, Bertha, which is housed the Archives of the History of American Psychology along with his other papers.

The construction of the Harlow Primate Laboratory is detailed in Clara Mears Harlow, ed., *Learning to Love: The Selected Papers of H. F. Harlow* (New York: Praeger, 1986); and in Harry Harlow, "Birth of the Surrogate Mother," *Discovery Processes in Modern Biology,* ed. W. M. Klemm (Huntington, 1977); and in Harlow's unpublished memoir.

Chapter Four: The Curiosity Box

L. R. Cooper and H. F. Harlow, "A Cebus Monkey's Use of a Stick As a Weapon," *Psychological Reports* 8 (1961): 418. Discussion of capuchin tool use in H. F. Harlow, "Primate Learning," in *Comparative Psychology,* ed. C. P. Stone, 3d ed. (New York, Prentice Hall, 1951), chapter 7. Also in Harry F. Harlow, "The Brain and Learned Behavior," *Computers and Automation,* vol. 4, no. 10 (October 1955).

Kohler cited in C. James Goodwin, *A History of Modern Psychology* (New York: J. Wiley, 1999). Maslow comment from Edward Hoffman, *The Right to be Human: A Biography of Abraham Maslow* (Los Angeles: J. P. Tarcher, 1988).

Kurt Goldstein in Ernest R. Hilgard, *Psychology in America: A Historical Survey* (San Diego: Harcourt Brace Jovanovich, 1987); in Harry F. Harlow, John P. Gluck, and Stephen J. Suomi, "Generalization of Behavioral Data Between Nonhuman and Human Animals," *American Psychologist,* vol. 27, no. 8 (August 1972); in Harry F. Harlow, "Mice, Monkeys, Men and Motives," *Psychological Review,* vol. 60 (1953); in Harry F. Harlow, "The Formation of Learning Sets," *Psychological Review,* vol. 56 (1949); and in Harry F. Harlow, "The Evolution of Learning," in Anne Roe and George Gaylord Simpson, eds., *Behavior and Evolution* (Yale University Press, 1958).

Thorndike in Harlow, "Mice, Monkeys, Men and Motives"; in Goodwin, *History of Modern Psychology;* in Hilgard, *Psychology in America;* and in Duane M. Rumbaugh, "The Psychology of Harry F. Harlow: A Bridge from Radical to Rational Behaviorism," *Philosophical Psychology,* vol. 10, no. 2 (1997). B. F. Skinner in his *The Shaping of a Behaviorist* (New York: Alfred A. Knopf, 1979), and his "Superstition in the Pigeon," *Journal of Experimental Psychology* 38 (1948), and in Hilgard and Rumbaugh.

Harry Harlow's comment on the Watsonian scourge is taken from his "Mice, Monkeys, Men and Motives." He further discusses Watson and B. F. Skinner in a speech on William James, "William James and Instinct Theory," American Psychological Association, September 4, 1967.

Clark Hull is profiled in Hilgard. He and Spence are discussed in detail in William Verplanck's autobiographical recollections, which are published on his Web site at http://web.utk.edu/~wverplan/default.html.

The Wisconsin General Test Apparatus (WGTA) is first described in Harry F. Harlow and John A. Bromer, "A Test Apparatus for Monkeys," *Psychological Record* 2 (1938): 434–436; a more modern version is disclosed in John W. Davenport, Arnold S. Chamove, and Harry F. Harlow, "The Semi-Automatic Wisconsin General Test Apparatus," *Behavioral Research Methods and Instruments,* vol. 2, no. 3 (1970). Allan Schrier's WGTA license plate is discussed in correspondence archived at the Archives of the History of American Psychology.

The rat blitzkrieg problem is described in Harry Harlow, "Formation of Learning Sets" (paper presented to the annual convention of Midwest Psychological Association, St. Paul, Minnesota, 7 May 1948). The tests run on the WGTA are described in that paper (Harlow's presidential address to the MPA) and were named as one of the top one hundred neuroscience discoveries of the twentieth century by the University of Minnesota. I won't cite the hundreds of other WGTA papers that followed, but I do want to mention three specifically: M. M. Simpson and H. F. Harlow, "Solution By Rhesus Monkeys of a Non-Spatial Delayed Response to the Color of Form Attribute of a Single Stimulus (Wiegl Principle Delayed Reaction)," *Journal of Comparative Psychology,* vol. 37, no. 4 (August 1944); Harry F. and Margaret Kuenne Harlow, "Learning to Think," *Scientific American*, August 1949; and Louis E. Moon and Harry F. Harlow, "Analysis of Oddity Learning by Rhesus Monkeys," *Journal of Comparative and Physiological Psychology,* vol. 48, no. 3 (June 1953).

The story of the light-switching spider monkey is in the unpublished memoirs and in "Birth of the Surrogate Mother," *Discovery Processes in Modern Biology,* ed. W. M. Klemm (Huntington, N.Y.: R. E. Krieger, 1977).

Papers on the curiosity studies include the following: Harry F. Harlow, "The Formation of Learning Sets: Learning and Satiation of Response in Intrinsically Motivated Complex Puzzle Performance By Monkeys," *Journal of Comparative and Physiological Psychology* 43 (1950): 289–294; Harry F. Harlow, Margaret Kuenne Harlow, and Donald R. Meyer, "Learning Motivated By a Manipulation Drive," *Journal of Experimental Psychology,* vol. 40, no. 2 (April 1950); Robert A. Butler and Harry F. Harlow, "Discrimination Learning and Learning Sets to Visual Exploration Incentives," *Journal of General Psychology*, vol. 57 (1957).

Chapter Five: The Nature of Love

Research into affection and intelligence is discussed in Goldfarb's work (also in Chapter 1). Further details from Goldfarb's paper, "The Effects of Early Institutional Care on Adolescent Personality," *Journal of Experimental Education*, vol. 12, no. 2 (December 1943); Robert Karen, *Becoming Attached: Unfolding the Mystery of the Infant-Mother Bond and Its Impact on Later Life* (New York: Warner Books, 1994); in Joel Shurkin, *Terman's Kids: The Groundbreaking Study of How the Gifted Grow Up* (Boston: Little, Brown and Co., 1992); in Ernest R. Hilgard, *Psychology in America: A Historical Survey* (San Diego: Harcourt Brace Jovanovich, 1987); and in John A. Popplestone and Marion White McPherson, *An Illustrated History of American Psychology* (Akron, Ohio: University of Akron Press, 1994).

Terman's prediction about Harry's American Psychological Association presidency can be found in a 1946 letter, archived at Stanford. Harry's comments about God, learning, and love come from an unpublished paper. Clara's comments about Harry's behavior, including the argument in which Harry questioned their love for each other during the breakup of their marriage are taken from the documents filed in support of her divorce petition at the Dane County Circuit Court. The property breakdown also comes from those documents.

The correspondence between Paul Settlage and Abe Maslow is housed at the Archives of the History of American Psychology.

Background on Margaret Kuenne Harlow is based on an interview with her brother, written answers to questions by her daughter, Pamela Harlow, and comments made by her son, Jonathan Harlow, in Richard Dukelow, *The Alpha Males: An Early History of the Regional Primate Research Centers* (Lanham, Mass.: University Press of America, 1995). The letter about Pamela's birth is archived at Stanford University, the Lewis Terman files. Other descriptions of Margaret Harlow are based on interviews with former students and staff.

I relied on two books about Carl Rogers: Howard Kirschenbaum, *On Becoming Carl Rogers* (New York: Delacorte Press, 1979); and Richard I. Evans, *Carl Rogers: The Man and His Ideas* (New York: Dutton, 1975). Kirschenbaum's book includes, in full, Rogers's parting memo to the UW psychology department.

John P. Gluck's descriptions of Harry are again taken from his "Harry Harlow: Lessons on Explanations, Ideas and Mentorship," *American Journal of Primatology* 7 (1984): 139–146.

Chapter Six: The Perfect Mother

Harry Harlow's description of the problems of importing monkeys and the "ghastly diseases" endemic to India and the issues of starting a breeding colony in "Birth of the Surrogate Mother," *Discovery Processes in Modern Biology*, ed. W. M. Klemm (Huntington, N.Y.: R. E. Krieger, 1977).

Stone's comments on sleeping at the primate lab are from a note to Richard Dukelow, archived at the library of the Wisconsin Regional Primate Research Center. The note is written on the cover of his paper, W. H. Stone, W. F. Blatt, and K. P. Link, "Immunological Consequences of Feeding Cattle Serum to the Newborn of Various Species," *Research Bulletin*, vol. 3, no. 1 (1957).

The work of Mason and Blazek with the "Stone" monkeys is described in "The Monkeys Who Go to College," which looked at curiosity testing of those little monkeys. The article appeared in the *Saturday Evening Post*, 15 October 1955.

Alfred R. Wallace's encounter with the baby orangutan is described in Deborah Blum, *The Monkey Wars* (New York: Oxford University Press, 1994), 89.

The discussion of Van Wagenen's work is housed at the Archives of the History of American Psychology.

The airplane story of the surrogate mother appears in many places, including the Harlows's "Birth of the Surrogate Mother."

Skinner's experiment with his daughter, Debbie, is described in his autobiography, *The Shaping of a Behaviorist* (New York: Alfred A. Knopf, 1979).

The baby monkey who loved the blank ball head is described in Harry's famous speech, "The Nature of Love," given at the 66th annual convention of the American Psychological Association, Washington D.C., 31 August 1958, and reprinted in *The American Psychologist*, vol. 13., no. 12 (1958). His comments from that speech are discussed in the latter section of the chapter as well. The research was first published outside the psychology community in H. F. Harlow and R. R. Zimmerman, "Affectional Responses in the Infant Monkey," *Science* 130, no. 3373 (1959); it also appears in Harry F. Harlow, "The Development of Affectional Patterns in Infant Monkeys," in *Determinants of Infant Behavior*, ed. B. M. Foss (New York: John Wiley & Sons, 1959); and in Harry F. Harlow, "Love in Infant Monkeys," *Scientific American*, vol. 6, no. 200 (1959).

Mary Ainsworth's pioneering work is discussed in Robert Karen, *Becoming Attached: Unfolding the Mystery of the Infant-Mother Bond and Its Impact on Later Life* (New York: Warner Books, 1994), Sarah Blaffer Hrdy, *Mother Nature: A History of Mothers, Infants, and Natural Selection* (New York: Pantheon Books, 1999), and Jude Cassidy and Phillip R. Shaver, eds., *Handbook of Attachment* (New York: The Guildford Press, 1999). Ainsworth appears throughout the book but her story is summarized in Cassidy's opening chapter, "The Nature of the Child's Ties," pp. 3–20 in the *Handbook.* Bowlby's connection to Konrad Lorenz is detailed in Hrdy's book, *Mother Nature,* and Karen's *Becoming Attached.*

Bowlby's correspondence with Harry Harlow is archived by Helen LeRoy at the Harlow Primate Laboratory; LeRoy has written a thoughtful paper discussing the relationship, titled "John Bowlby and Harry Harlow: The Cross Fertilization of Attachment Behavior Theory."

Chapter Seven: Chains of Love

L. Joseph Stone, Henrietta T. Smith, and Lois B. Murphy, eds., *The Competent Infant: Research and Commentary* (New York: Basic Books, 1973).

Correspondence between Harry Harlow and Joseph Stone and Nancy Bayley is housed at the Archives of the History of American Psychology.

Hebb cited in Harry F. Harlow, "The Brain and Learned Behavior," *Computers and Automation,* vol. 4, no. 10 (October 1955). Hebb's correspondence with Harlow is archived at McGill University, but psychology historian, Steve Glickman, of the University of California-Berkeley and a former Hebb student, kindly provided me with copies. Hebb is considered by many psychologists to be the outstanding theorist of the mid-twentieth century; his theories about the effect of experience on neurons in the brain are considered classics of their time. He is only briefly a part of this particular story but in the larger sense of psychology history, he deserves much more credit.

The University of Wisconsin has archived hundreds of press clippings, and press releases from Harry Harlow's heyday in the 1960s in particular. To give a sense of the popular appeal of his work, I'd like to cite: John Kord Lageman, "What Monkeys Are Teaching Science About Children," *This Week,* 3 March 1963; "Can Mothers Be Replaced?" *Picture Magazine,* 26 July 1959; Clarissa Start, "Raising Baby Monkeys with Cloth Mother," *St. Louis Post Dispatch,* 3 May 1960.

The problems with the surrogate raised mothers are described in Leonard Engel, "The Troubled Monkeys of Madison," *New York Times,* January 29, 1961.

Early experiments in rat-handling are detailed in Seymour Levine, "A Further Study of Infantile Handling and Adult Avoidance Learning," *Journal of Personality* 25 (1956): 70–80; Seymour Levine and Leon S. Otis, "The Effects of Handling Before and After Weaning on the Resistance of Albino Rats to Later Deprivation," *Canadian Journal of Psychology* 12 (1958): 2; Seymour Levine and George W. Lewis, "Critical Period for Effects of Infantile Experience on Maturation of Stress Response," *Science* (1959): 129, 42–43; Theodore Schaefer, Jr., "Some Methodological Implications of the Research on 'Early Handling' in the Rat," in Grant Newton and Seymour Levine, eds., *Early Experience and Behavior: The Psychobiology of Development* (Springfield, Ill.: Charles C. Thomas, 1968).

Victor Denenberg's related studies appear, among other places, in Victor Denenberg and Robert Bell, "Critical Periods for the Effects of Infantile Experience on Adult Learning," *Science*, vol. 131 (1960); and in Victor Denenberg and John C. Morton, "Effects of Environmental Complexity and Social Groupings Upon Modification of Emotional Behavior," *Journal of Comparative and Physiological Psychology*, vol. 55, no. 2 (1955).

Both "hot mamma" and rocking surrogates are discussed in Harry F. Harlow and Stephen J. Suomi, "The Nature of Love—Simplified," *American Psychologist*, vol. 25, no. 2 (February 1970). The studies continued, as reported in C. M. Baysinger, P. E. Plubell, and H. F. Harlow, "A Variable Temperature Surrogate Mother for Studying Attachment in Infant Monkeys," *Behavioral Research and Methods*, vol. 5, no. 3 (1973).

William Mason and Gershon Berkson's work with motion is described in "Effects of Maternal Mobility on the Development of Rocking and Other Behaviors in Rhesus Monkeys: A Study with Artificial Mothers," *Developmental Psychobiology* 8, no. 3 (1975): 197–211; in M. V. Neal, "Vestibular Stimulation and Developmental Behavior of the Small Premature Infant," *Nursing Research Report*, American Nurses Foundation, vol. 3, no. 1 (March 1968); and in further detail, along with later studies including the rocking horse surrogates, in William A. Mason, "Social Experience and Primate Cognitive Development," *The Development of Behavior: Comparative Evolutionary Aspects*, ed. Gordon Burghardt and Mark Bekoff (New York: Garland STPM Press, 1978); William A. Mason, "Maternal Attributes and Primate Cognitive Development," in *Human Ethology: Claims and Limits of New Discipline*, ed. M. Von Cranach, K. Foppa, W. Lepenies, and D. Floog (Cambridge: Cambridge University Press, 1979).

Harry Harlow's comment on the limits of the surrogate mother cited in Engel, "Troubled Monkeys." The role of the mother as socializer is further discussed in "The Effect of Rearing Conditions on Behavior," a presentation by Harry F. Harlow and Margaret K. Harlow to a forum of the Menninger School of Psychiatry, December 4, 1961. The issue of mothers who keep their children too close to home is discussed in G. W. Moller, H. F. Harlow, and G. D. Mitchell, "Factors Affecting Agonistic Communication in Rhesus Monkeys," *Behavior,* vol. 31 (1968).

Mother-infant relationship studies: L. A. Rosenblum and H. F. Harlow, "Approach-Avoidance Conflict in the Mother Surrogate Situation," *Psychological Reports,* vol. 12 (1963); and L. A. Rosenblum and H. F. Harlow, "Generalization of Affectional Responses in Rhesus Monkeys," *Perceptual and Motor Skills,* vol. 16 (1963).

Robert Hinde's perspective is found in Patrick Bateson, ed., *The Development and Integration of Behavior: Essays in Honor of Robert Hinde* (Cambridge: Cambridge University Press, 1991); and Robert Hinde, *Individuals, Culture and Relationships* (Cambridge: Cambridge University Press, 1987). I read both books on recommendation of Professor Hinde and found them an enlightening look at the science of relationships, a theme that runs through this book.

Different primate parenting styles in M. W. Andrews and L. A. Rosenblum, "Assessment of Attachment in Differentially Reared Infant Monkeys—Response to Separation and a Novel Environment," *Journal of Comparative Psychology,* vol. 107, no. 1 (March 1993).

The study of peer relationship is addressed in A. S. Chamove, L. A. Rosenburg, and H. F. Harlow, "Monkeys Raised Only with Peers: A Pilot Study," *Animal Behavior* 21, no. 2 (1973): 316–325; Stephen J. Suomi and Harry F. Harlow, "The Role and Reason of Peer Relationships in Rhesus Monkeys," in *Friendship and Peer Relationships,* ed. Lewis Rosenblum (New York: John Wiley and Sons, 1975); Harry F. Harlow, "Age-Mate or Peer Affectional System," *Advances in the Study of Behavior,* vol. 2, 1969; and Stephen J. Suomi, "Peers, Play and Primary Prevention in Primates," in *Proceedings of the Third Conference on Primary Prevention of Psychopathology* (Hanover N.H.: University Press of New England, 1979).

The first paper published on the studies of monkey family support systems was "Nuclear Family Apparatus," Margaret K. Harlow, *Behavioral Research Methods and Instruments,* vol. 3, no. 6 (1971). Many of the other papers appeared after Margaret Harlow's death. I relied particularly on G. C. Ruppenthal, M. K. Harlow, C. D. Eisele, H. F. Harlow, and S. J. Suomi, "Development of Peer Interactions of Monkeys Reared in a Nuclear-Family Environment," *Child Development* 45 (1974): 670–682. I also found the discussion in Harry F. Harlow and Clara Mears Harlow, eds., *Learning to Love: The Selected Papers of H. F. Harlow* (New York: Praeger, 1986), enlightening.

The monster mothers are described in Stone, Smith, and Murphy, *Competent Infant.*

Chapter Eight: The Baby in the Box

Doggerel from Harry Harlow, "The Nature of Love," presidential address, 66th annual convention of the American Psychological Association, Washington, D.C., August 31, 1958.

Further description of nuclear family actions from G. C. Ruppenthal, M. K. Harlow, C. D. Eisele, H. F. Harlow, and S. J. Suomi, "Development of Peer Interactions of Monkeys Reared in a Nuclear-Family Environment," *Child Development,* vol. 45, (1974), and in Harry F. Harlow and Clara Mears, *The Human Model: Primate Perspectives*, New York, John Wiley & Sons, 1979.

The "Hell of Loneliness" chapter is in Harlow and Mears, *The Human Model.*

Depression and isolation studies include: "The Effect of Total Social Deprivation on the Development of Monkey Behavior," in *Psychiatric Research Report,* vol. 19, American Psychiatric Association (December 1964); "Total Social Isolation in Monkeys," *Proceedings of the National Academy of Sciences,* vol. 54, no. 1 (1965); Harry F. Harlow and Billy Seay, "Mothering in Motherless Mother Monkeys," *The British Journal of Social Psychiatry,* vol. 1, no. 1 (1966).

Harry F. Harlow and Stephen J. Suomi, "Production of Depressive Behaviors in Young Monkeys," *Journal of Autism and Childhood Schizophrenia* 1, no. 3 (1971): 246–255; "Depressive Behavior in Young Monkeys Subjected to Vertical Chamber Confinement," *Journal of Comparative and Physiological Psychology* 180, no. 1 (1972): 11–18; Harry F. Harlow, Philip E. Plubell, and Craig M. Baysinger, "Induction of Psychological Death in Rhesus Monkeys," *Journal of Autism and Childhood Schizophrenia* 3, no. 4 (1973): 299–307; Stephen J. Suomi, Mary L. Collins, and Harry F. Harlow, "Effects of Permanent Separation from Mother on Rhesus Monkeys," *Developmental Psychology,* vol. 9, no. 3 (1979); "Induced Depression in Monkeys," *Behavioral Biology* 12 (1974): 273–296; Stephen J. Suomi, Carol D. Eisele, Sharon A. Grady, and Harry F. Harlow, "Depressive Behavior in Adult Monkeys Following Separation from Family Environment," *Journal of Abnormal Psychology* 84, no. 5 (1975): 576–578.

Peer therapy work includes: H. F. Harlow, M. K. Harlow, and S. J. Suomi, "From Thought to Therapy: Lessons from a Primate Laboratory," *American Scientist* (September-October 1971); Stephen J. Suomi, Harry F.

Harlow, and Melinda A. Novak, "Reversal of Social Deficits Produced by Isolation Rearing of Monkeys," *Journal of Human Evolution* 3 (1974): 527–534; Harry F. Harlow and Stephen J. Suomi, "Social Recovery By Isolation-Reared Monkeys," *Proceedings of the National Academy of Sciences* 68, no. 7 (1971); Harry F. Harlow and Melinda A. Novak, "Psychopathological Perspectives," *Perspectives in Biology and Medicine*, vol. 16, no. 3 (1973); "Social Recovery of Monkeys Isolated for the First Year of Life," *Developmental Psychology* 11, no. 4 (1975).

Harry's comment on the university's handling of Margaret Harlow appears in Carol Tavris, "Harry, You Are Going to Go Down in History As the Father of the Cloth Mother," *Psychology Today* (April 1973).

Chapter Nine: Cold Hearts and Warm Shoulders

Bruno Bettelheim wrote about refrigerator mothers and their relationship to Harry Harlow's surrogates in *The Empty Fortress: Infantile Autism and the Birth of Self* (New York: The Free Press, 1967).

Harry Harlow's review of the book: H. F. Harlow, "A Brief Look At Autistic Children," *Psychiatry & Social Science Review* 3, no. 1 (January 1969): 27–29.

Sarah Blaffer Hrdy writes about scientific attitudes toward women scientists in *Mother Nature: A History of Mothers, Infants, and Natural Selection* (New York: Pantheon Books, 1999).

Diane E. Eyers, *Mother-Infant Bonding* (New Haven and London: Yale University Press, 1992).

The Maggie story appears in Harry F. Harlow, "Birth of the Surrogate Mother," *Discovery Processes in Modern Biology*, ed. W. M. Klemm (Huntington, N.Y.: R. E. Krieger, 1977).

Jane Glascock's letter about Harry Harlow's appearance at the University of Washington-Seattle is housed at the Archives of the History of American Psychology.

The description of Clara Mears Harlow's life in between her two marriages to Harry is outlined in her file with the Lewis Terman gifted project archive at Stanford University, and also based on interviews with her two sons, Robert Israel and Richard Potter.

Harry Harlow's correspondence with William Verplanck is housed at the Archives of the History of American Psychology.

The quotes about Wisconsin weather and retirement are in "Behavioral Giant Not Going to Seed," *The Capital Times*, Madison, Wisconsin, August 3, 1978.

Sears's letter is archived in the Lewis Terman gifted project files at Stanford. The letter to William Mason courtesy of Mason's personal files.

Harlow's doggerel from his days in Arizona is courtesy of Robert Israel's private collection.

Clara Harlow's correspondence with Sears archived in the Lewis Terman gifted project files.

Chapter Ten: Love Lessons

A review of the way human infants watch, read, and respond to their mother's faces, including the virtual cliff experiment, can be found in the writings of Edward Z. Tronick: Jeffrey Cohn and Edward Tronick, "Specificity of Responses to Mother's Affective Behavior," *Journal of the American Academy of Child Adolescent Psychiatry* 28, no. 2 (1989): 242–248; Edward Z. Tronick, "Emotions and Emotional Communication in Infants," *American Psychologist* 44, no. 2 (February 1989): 112–119; L. Murray and P. Cooper, eds., "Depressed Mothers and Infants: Failure to Form Dyadic States of Consciousness" in *Postpartum Depression and Child Development* (New York: Guilford Press, 1997); Edward Z. Tronick and Andrew Gianino, "Interactive Mismatch and Repair: Challenges to the Coping Infant," *Zero to Three,* vol. 6, no. 3 (February 1986); "Dyadically Expanded States of Consciousness and the Process of Therapeutic Change," *Infant Mental Health Journal* 19, no. 3 (1998): 290–299.

Judith Rich Harris, *The Nurture Assumption: Why Children Turn Out the Way They Do* (New York: The Free Press, 1998).

Harry F. Harlow and Margaret K. Harlow, "Learning to Love," *American Scientist* 54, no. 3 (1966): 244–272. This was also the title of a book coedited by Harry and Clara Harlow.

See previous Bowlby citations, also Mary Salter Ainsworth and John Bowlby, "An Ethological Approach to Personality Development," *American Psychologist,* vol. 46, no. 4 (April 1991): 333–341; and *Deprivation of Maternal Care* (World Health Organization [WHO] report, Geneva, 1962).

Marshall H. Klaus and John H. Kennell, *Bonding: The Beginnings of Parent-Infant Attachment,* rev. ed. (New York: New American Library, 1983).

Marshall H. Klaus, John H. Kennell, and Phyllis H. Klaus, *Bonding: Building the Foundations of Secure Attachment and Independence,* first paperback ed. (Reading, Mass.: Addison-Wesley, 1996.)

Meredith F. Small, *Our Babies, Ourselves: How Biology and Culture Shape the Way We Parent* (New York: Anchor Books, 1998); *Kids: How Bi-*

ology and Culture Shape the Way We Raise Our Children (New York: Doubleday, 2001).

The NICHD study of childcare discussed in Deborah Blum, *Sex on the Brain: The Biological Differences Between Men and Women* (New York: Viking, 1997). One of the best perspectives on day care is Ellen Ruppel Shell, *A Child's Place: A Year in the Life of a Daycare Center* (Boston: Little, Brown and Co., 1992).

Suomi's work on mothering styles is found in Jude Cassidy and Phillip R. Shaver, eds., "Attachment in Rhesus Monkeys," *Handbook of Attachment* (New York: The Guildford Press, 1999).

Leonard Rosenblum's look at "aunted" monkeys is cited in L. T. Nash and R. L. Wheeler, "Mother-Infant Relationships in Non-Human Primates," in Hiram E. Fitzgerald, John A. Mullins, and Patricia Gage, eds., *Child Nurturance,* vol. 3 (New York, Plenum Press, 1982). He is co-editor with Michael Lewis of two books that further explore such relationships: Michael Lewis and Leonard A. Rosenblum, eds., *The Effect of the Infant on Its Caregiver* (New York: John Wiley & Sons, 1974); and *The Child and Its Family* (New York: Plenum Press, 1979).

The behavior of titi monkeys appears in Sally P. Mendoza and William A. Mason, "Parental Division of Labor and Differentiation of Attachments in a Monogamous Primate," *Animal Behavior,* vol. 34 (1986).

Cotton top tamarin society is discussed in Charles T. Snowdon, "Infant Care in Cooperatively Breeding Species," *Advances in the Study of Behavior,* vol. 25, 1996; and Gretchen G. Achenbach and Charles T. Snowdon, "Response to Sibling Birth in Juvenile Cotton Top Tamarins," *Behaviour,* vol. 135, no. 7 (1998).

Myron A. Hofer, "Infant Separation Responses and the Maternal Role," *Biological Psychiatry,* vol. 10, no. 2 (1975).

Saul M. Schanberg and Tiffany M. Field, "Sensory Deprivation Stress and Supplemental Stimulation in the Rat Pup and Preterm Human," *Child Development,* vol. 58 (1987); S. M. Schanberg, "Medicine: Different Strokes," *Scientific American* (September 1989): 34; "Touch: A Biological Regulator of Growth and Development in the Neonate," *Verhaltenstherapie,* vol. 3, Suppl. 15 (1993); Daniel Goleman, "The Experience of Touch: Research Points to a Critical Role," *New York Times,* 2 February 1988.

Sapolsky's discussion of rat stress experiments is found in *Why Zebras Don't Get Ulcers* (W. H. Freeman and Co., 1994). The studies with Michael Meaney are also discussed in that book. Further information on Meaney's research comes from his presentation at the February 2001 meeting of the American Association for Advancement of Science in San Francisco and from discussion with another of his research colleagues, Paul Plotsky, at Emory University in Atlanta.

Martin H. Teicher, "Scars That Won't Heal: The Neurobiology of Child Abuse," *Scientific American,* vol. 286, no. 3 (March 2002).

Bruce Perry quoted in Deborah Blum, "Attention Deficit," *Mother Jones* (January/February 1999). Information on face-reading skills from Deborah Blum, "Let's Face It," *Psychology Today,* vol. 31, no. 5 (September/October 1998).

Harry's comment about love and modesty in a letter to William Verplanck, housed at the Archives of the History of American Psychology in Akron.

Epilogue: Extreme Love

Harry Harlow's discussion of the aspects of love and the importance of primate research in answering questions of child abuse in "Behavioral Giant Not 'Going to Seed,'" *The Capital Times,* Madison, Wisconsin, August 3, 1978.

The proposed NIH experiment on isolating children is described in a textbook: Harry F. Harlow, James L. McGaugh, and Richard F. Thompson, eds., *Psychology* (San Francisco: Albion Publishing Company, 1971).

Thomas Lewis, Fari Amini, and Richard Lannon, *A General Theory of Love* (New York: Random House, 2000).

Sapolsky quotes on animal research from his *Why Zebras Don't Get Ulcers* (W. H. Freeman and Co., 1994).

Harlow comments on his perspective on monkeys from Robert Bonin, "Harry Harlow Has Spent a Lifetime Studying Monkeys—Which Doesn't Mean He Likes Them," *Milwaukee Journal,* October 28, 1973.

Mason's comment from Deborah Blum, *The Monkey Wars* (New York: Oxford University Press, 1994).

Martin Stephens's survey of isolation experiments is called "Maternal Deprivation Experiments in Psychology." It was published in 1986 for the American Anti-Vivisection Society and the New England Anti-Vivisection Society.

The descriptions of other animal research and the brief history of the animal rights movement, including comments from Christine Stevens, a discussion of the Silver Spring monkeys, and the letter to Gig Levine, come from Blum, *The Monkey Wars.* Gluck's comment is from an interview for this book.

Index